KARAWEBER

Introduction to BIOLOGY

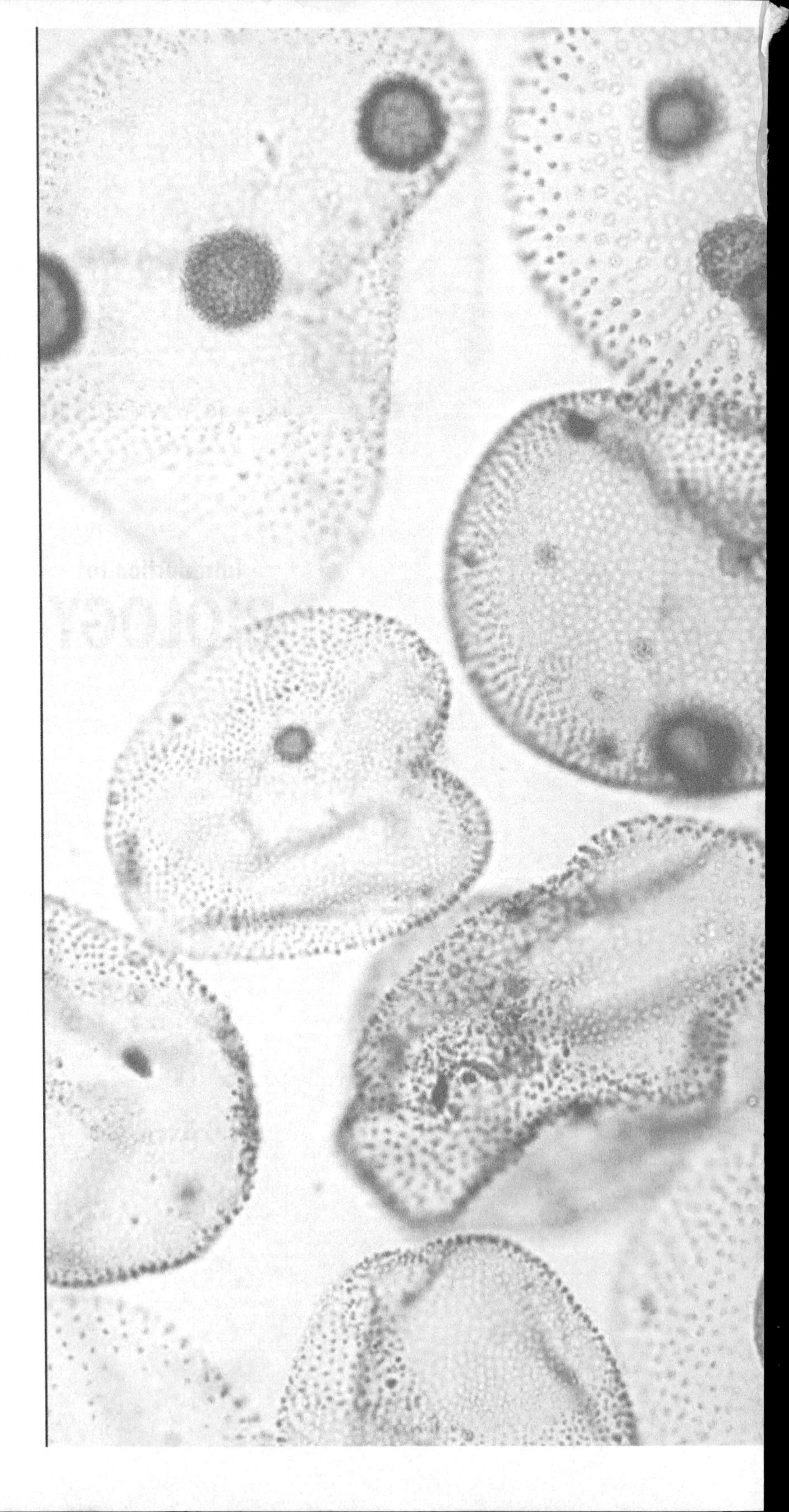

PREFACE

Biology is a rapidly developing branch of science. The major advances that are made, continuously affect our life on earth. Some of these important advances are included here.

The results of a recent survey on the attitudes to existing literature available to high school students showed that many were unhappy with the material used in teaching and learning. Those questioned identified a lack of the following; accompanying supplementary material to main text books, current information on new developments, clear figures and diagrams and insufficient attention to design and planning of experiments.

This book aims to improve the level of understanding of modern biology by inclusion of the following; main texts, figures and illustrations, extensive questions, articles and experiments.

Each topic is well illustrated with figures and graphs to ease understanding. Supplementary material in the form of posters, transparencies and cassettes will shortly be available.

Profiles on common diseases are included in each chapter to inform, generate further interest and encourage students to explore the subject further. The 'Read me' articles supply up-to-date information on important issues related to each unit but outside the requirements of the current curriculum.

It is the intention and hope of the authors that the contents of this book will help to bridge the current gap in the field of biology at this level.

This book has been carefully reviewed and the language is considered suitable for students for whom English is a second language.

All errors in the previous edition have been corrected.

CONTENTS

I. SCIENCE AND SCIENTIFIC METHOD

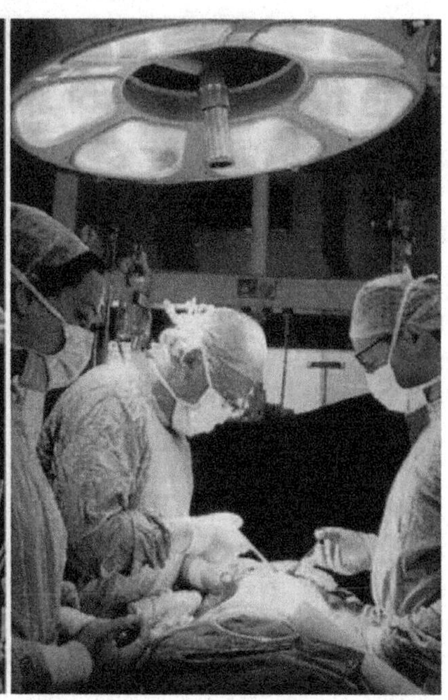

A. WHAT IS SCIENCE

[illegible] Humans first tried to answer these questions with magic or legends, but later explored the scientific approach. Indeed, the concept of science can be related to everything that people experience throughout their life. In a similar manner, [illegible]

A scientist is one who develops methods to unravel the mysteries of nature and the secrets of life.

The first step that must be taken by scientists who have dedicated themselves to science is to examine possible ways of being useful to humanity.

The foremost factor in forming a proper conclusion in any scientific research is to be unbiased and to generate a hypothesis (educated guess) from controlled experiments in the laboratory and observations in nature.

Also, it must be accepted that it is not very easy to get suitable results in scientific research. A scientist should have curiosity. Patience, determination, and ambition are some other characteristics of a modern scientist.

One method, called the "**scientific method**", has been commonly adopted and is used by modern scientists. There are a series of steps in the scientific method. read hmw

B. How does a scientific problem arise?

For all events in life, whether people are aware of them or not, there are questions waiting to be considered and reasonable answers to be found. For example, how a combination of an egg and sperm develops into an embryo in the uterus, or through what stages a plant passes as it develops from a seed to a mature tree. By asking how and why, we are eventually surrounded

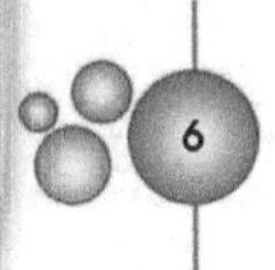

by questions concerning every aspect of daily life. There are many facts of life that we have observed, such as the falling of an apple from a tree. As a scientist, Newton questioned this event and had a scientific method to use to learn the reason for it.

In scientific research, usually the methods of induction and deduction are used.

Deductive reasoning is a method in which preliminary information is used. This information is taken as reference, and the solution is determined from it. For example: Let's accept that all birds have wings. Because we know that sparrows are also birds, we can say that sparrows have wings too.

However, in the inductive reasoning method, special observations, such as the fall of a free object, are the first steps in research. From observations of this event we may get a result or we may discover a new fact. With this method, all observations are gathered and there is an attempt to solve the problem. Most hypotheses and theories are the products of the inductive reasoning method. The generalization of some results to all events is the weakest point of this method.

C. THE STAGES OF

There is a guide, called the scientific method, by which a scientist proceeds with an investigation into an area of logic. There are various stages in the method.

1.

as to particular or she is going to research. Then, he or she should define the problem and ask the question which he/she

2.

Qualitative observations are made by using the five senses. There is no need for data-gathering instruments. In these types of observations, the thoughts of the observer are acceptable criteria. This might be the color of an apple or the flavor of an orange. These observations are relative to the observer, and are subjective.

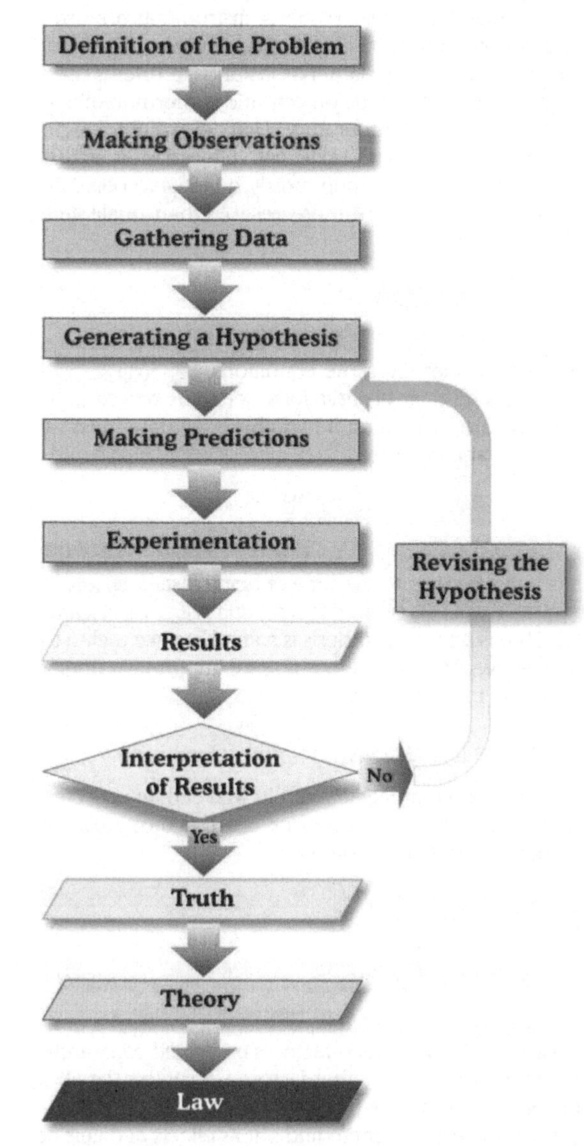

Figure 1.1.: *The flow chart of the scientific method. As seen in the figure, if the hypothesis can't be proven by the experiments, it should be revised, and the next steps should be repeated. When the hypothesis is supported by experiments and many scientists, it will become theory. When all scientists confirm it, it is called law.*

In quantitative observations, instruments are used to determine and obtain data from events. There are criteria for gathering data. For instance, a ruler is used to find the length of an object, and a thermometer is used to measure temperature. Units of weight, length and temperature are universal. Objective scientists gather their data by using mostly quantitave observations because they are more reliable than qualitative observations.

3. Gathering Data

that are always the same under the same conditions, After making their observations, scientists record their data. Before generating a hypothesis, all data must be gathered and organized.

4.

It is temporary up to the time of being tested. scientific , or discarded. That is why a hypothesis is something like a clue to a detective. Such a clue has no real importance until it is proven in court.

5. Making Predictions

Predictions are logical results that can be expected from a hypothesis. Predictions are the most reliable methods of testing hypotheses.

6.

validity of a a controlled Each factor is examined separately. In the experiment, all other factors, except this variable factor, are fixed. The variable factor is varied in the steps of the experiment to find out its effects at different steps.

For example, to understand the effects of magnesium (Mg) in the development of plants, at least two plants from the same species must be used. To one of them, a normal amount of magnesium, to the other, a variable amount of magnesium must be given. To observe differences in the growth of the plants, the effect of other factors that may affect plant development, such as temperature, humidity, water, and oxygen, must be kept constant. Any little mistake in the experiment may give an inaccurate result.

7.

If results and predictions contradict each other, then the hypothesis must be revised.

8.

found to be largely repeated s,

It is always possible to disprove a theory. For example, Dalton's atomic theory stated that the atom is the smallest unit of matter. Today this theory has no validity.

9.

For instance, Mendel's laws are indisputable because they are accepted by all scientist and are perfectly and consistently confirmed.

Important !!! READ

D. AN EXAMPLE OF THE SCIENTIFIC METHOD

The scientific method mentioned before is applied in much research by most scientists. One of the experiments done by Pasteur, who obeyed the rules of the scientific method, will make it possible to better understand this method.

At the time of Pasteur, one of the most controversial problems was **"the origin of life"**. Almost all scientists believed in the abiogenesis hypothesis, which states that **"living things arise from non-living things"**, such as mud turning into fish, or worms and insects forming from a piece of meat.

The steps of Pasteur's experiments can be summarized as follows.

– What is the origin of life? (The problem)

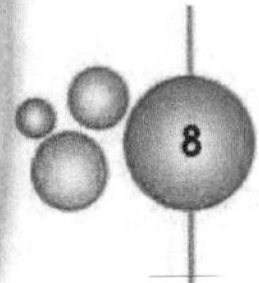

– Uncovered food (organic substances) spoils - (observation)

– Non-living things can't give rise to living organisms (hypothesis).

– Bacteria and microorganisms can be found on dust particles and carried by the air (prediction).

Experimentation to test the validity of the predictions.

An experiment is designed to test the hypothesis.

A solution, including sugar and yeast cells, is put into a flask and boiled. Then a curved tube is added to the top of the flask.

Through this curved tube a small amount of air could pass, but dust particles were trapped. Once more the solution was boiled. The flask stayed outside for a long time, but no living organisms were noticed in the solution.

As an experimental control, a sample was taken from the same solution and put into a flask without any cover or curved tube. The solution was boiled. After a period of time, microorganisms appeared in the solution (examination of the hypothesis).

As a result of this experiment, it was proved that non-living things do not spontenously change into living things (TRUTH). Now, all scientists agree on this truth (Law).

a)

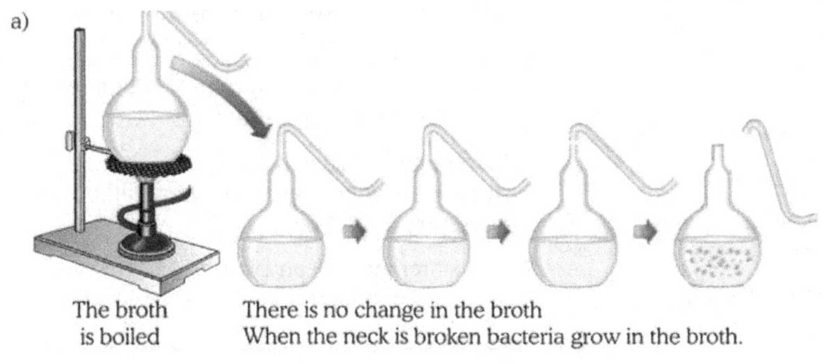
The broth
is boiled
There is no change in the broth
When the neck is broken bacteria grow in the broth.

b)

Unboiled broth

There is no change in the broth

c)

The broth
is boiled
After a while the development of
microorganisms is observed in the broth

Figure 1.2.:

Pasteur's Experiment :

Pasteur put clear broth into two flasks with curved necks, and into one without a curved neck. The curved neck allowed oxygen in but kept bacteria in the air out. Pasteur boiled the broth in the curved-neck flask (a) and in the flask without a neck (c) to kill bacteria in the broth.

As a control, the broth in the other curved-neck flask (b) wasn't boiled.

After a few days, the unboiled broth and the broth in the flask without a curved neck became cloudy. On the other hand, the boiled broth in the curved-neck flask remained clear.

Later he broke the neck of the curved-neck flask that contained clear broth. In a few days the broth became cloudy as bacteria grew.

He concluded that bacteria don't spontaneously grow from broth. New bacteria appeared only when living bacteria were already present.

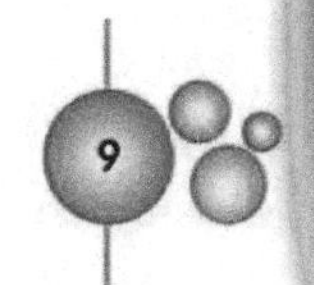

SELF CHECK

SCIENCE

A. Key Terms

Science	Deductive reasoning
Scienctific method	Inductive reasoning
Controlled Experiment	Natural law
Theory	Hypothesis
Spontaneous generation	Qualitative observation
Quantitave observation	

B. Review Questions

1. Briefly define the term science.
2. List the steps of the scientific method.
3. What does controlled experiment mean?
4. What should be done after repeated experiments show the hypothesis is wrong?
5. What were the conclusions from Pasteur's experiment?

C. True or False

1. ☐ Aristotle was the first scientist to name organisms with binomial nomenclature.
2. ☐ A controlled experiment involves performing at least two tests that are identical in every respect except one.
3. ☐ One factor is called a variable.
4. ☐ Pasteur concluded that bacteria grew spontaneously from the broth in his experiment.

D. Matching

a. Hypothesis ( ) Proven hypothesis with experiments.

b. Data ( ) Temporary solution to the problem.

c. Deduction ( ) Results of observations.

d. Induction ( ) Preliminary information is used.

e. Theory ( ) Generalization is the weakest point of this method.

E. Fill in the blank

1. In the 18th century, most scientists believed the .... hypothesis of the origin of life.
2. There is no need for any instruments with .... observations. They can be made with the five senses. They are subjective.
3. One method, called ...., is commonly adopted and used by modern scientists to solve scientific problems.
4. In scientific research, the .... should be generated first after gathering the necessary data.
5. The statement "bean plant is 11 cm long" is a .... observation.

F. Multiple choice

1- Which of the following has the highest possibility of being disproved ?

A) Theory B) Data C) Hypothesis D) Law E) Truth

2- What is the first temporary solution for a problem?

A) Theory
B) Law
C) Observation
D) Experiment
E) Hypothesis

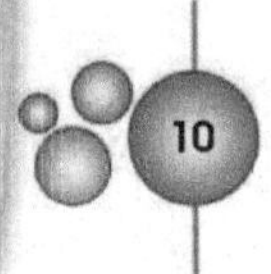

II. BIOLOGICAL SCIENCE

In this century, many scientific investigations have been conducted. As a branch of science, biology is also a developing process. Compared to previous generations, we know much more about biology, because of the acquired knowledge and experience transmitted by books and other resources.

For many years, the true value of biology wasn't recognized. But nowadays, the importance of biology is much better understood and it has started to receive the recognition it deserves. What made biology more important, or what caused its real importance to be better understood? Probably, the answer is the latest investigations going on around us: AIDS, its characteristics and how to avoid it; the growth of many different types of plants and animals with the help of selection; cloning organisms; the mystery of DNA and the gene map; air and water pollution and their effects on organisms; etc. All these things directly related to biology have made people think much more about biology.

Today it is accepted that the more biology is used in a country (agriculture, genetics, medicine, production of biological weapons) the more that country is developed.

As a science, biology tries to support students with knowledge that will help them throughout their lives. In biology, students learn their own anatomy, the stages of their existence, the changes going on during development, etc.

Biology is the science that studies nature, living things, the environment they live in, and interactions between organisms and the environment. As a result, biology can be called **"the science of life"** or **"the study of life"**.

GENETICS: The study of inheritance

ANATOMY: Study of internal structures of living things

EVOLUTIONARY BIOLOGY: Study of the origin and history of life

ZOOLOGY: The study of animals

PHYSIOLOGY: Study of functions of living systems

BOTANY: The study of plants

ETHOLOGY: Study of animal behavior

EMBRYOLOGY: Study of development in embryos

TAXONOMY: Study of classification, identification and naming of species

HISTOLOGY: Study of tissues

BIOLOGY

MORPHOLOGY: Study of outward appearance of organisms

CYTOLOGY: (Cell Biology) Study of cell structures and functions

ECOLOGY: Study of interactions between organisms and their environment

MICROBIOLOGY: The study of microscopic life

MYCOLOGY: Study of fungi

MOLECULAR BIOLOGY: Study of life on the level of molecular activity

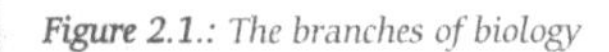

Figure 2.1.: The branches of biology

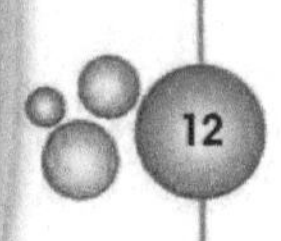

A. THE BRANCHES OF BIOLOGY

One of the basic aims of biology is to offer useful and technically feasible applications of the discoveries made by it for the benefit of the natural world. Today, biology's field of study is so broad that it has been divided into many sub-branches. As biologists open up the world of biology through research, new branches continue to be discovered.

1. Zoology

the animal kingdom has many members, [illegible]o [illegible]ologists sub-

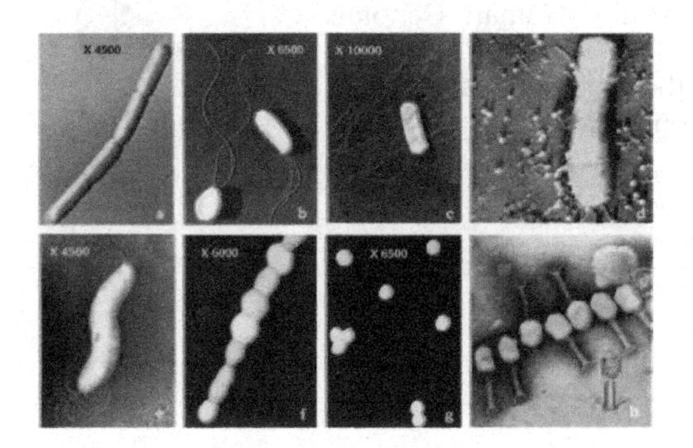

Microbiology : concentrates on <u>microorganisms</u>

Parasitology : the study of <u>parasites</u>

Mycology : the science of <u>fungi</u>

2. Botany

3. Anatomy

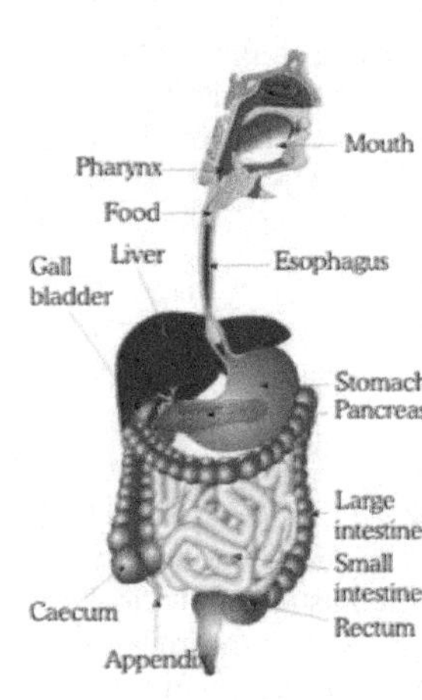

Anatomy is the study of the internal organs of the body (kidney, stomach, heart, bones, and so on). Anatomy analyzes organs and the differences in their structures. It is also concerned with the functions and structures of each organ. Anatomy is

s.

4. Morphology

5. Histology

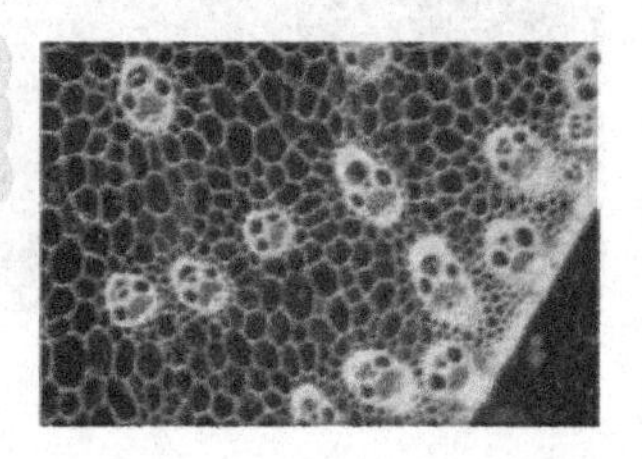

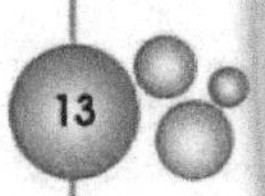

6. Physiology

Physiology is the integration of anatomy and histology. Physiology is concerned with how tissues, organs and organ systems function.

7. Embryology

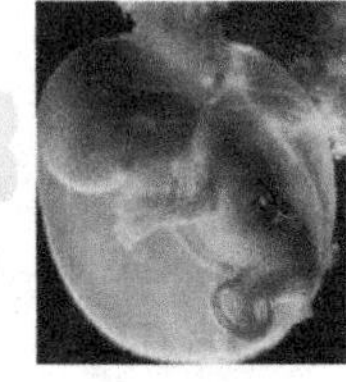

8. Cytology: "Cell Biology"

Cytologists analyze the structure and metabolism of cells, particularly when their functions are disrupted by disease.

9. Taxonomy

Taxonomists try to sort organisms into logical groups according to similarities in their body structures and functions. Organisms with the same characteristics are put together in the same groups.

10. Ecology

Ecology examines the changes occuring in all (living and non-living) environments and predicts possible outcomes of those changes in the future.

11. Genetics

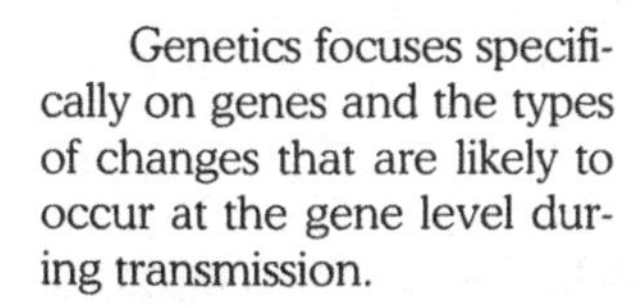

Genetics focuses specifically on genes and the types of changes that are likely to occur at the gene level during transmission.

Currently, genetics is at the forefront of biology and the term **"genetic engineering"** is becoming familiar to all. With new developments and discoveries in genetics, scientists even manipulate genes. They can now cure some genetic disorders, or at least keep them under control. Some other areas of particular interest to geneticists are important industrial plants, such as **cereals, sunflowers, vegetables and fruits**. Genes that control large fruit size, resistance to infection, etc. are of special importance.

Clonning of mammals and the Human Genome Project I and II are some other popular studies in genetics.

SELF CHECK

BIOLOGY

A. Key Terms

Biology	Botany
Ecology	Physiology
Anatomy	Morphology
Cytology	Histology
Zoology	Genetics
Embryology	Taxonomy

B. Review Questions

1. What is the goal of biology?
2. What is genetics?
3. Why do we need to know biology?
4. List the applications of biology in your daily life.
5. List some major branches of biology.

C. True or False

1. ☐ Botany is the study of the plant kingdom.
2. ☐ Morphology is the science of functions.
3. ☐ Genetics deals with inheritance between generations.
4. ☐ Cytology (Cell Biology) concentrates on tissues.
5. ☐ Viruses are the subject of virology.

D. Matching

a. Anatomy ( ) Science of functions

b. Zoology ( ) Deals with organisms from zygote to birth

c. Embryology ( ) Concerned with animals

d. Physiology ( ) Studies the internal structure of organisms.

e. Morphology ( ) Studies the external features of organisms.

E. Fill in the blank

1. .... is the naming, description and classification of living organisms.
2. Tissues are the subject matter of ....
3. .... a sub-branch of zoology, is concerned with insects.
4. Embryology is the study of living things from .... to ....
5. .... is the science of bacteria.

F. Multiple choice

1. Which of the following does study of microscopic organisms ?

A) Anatomy B) Zoology C) Entomology d) Microbiology e) Histology

2. Which one of the followings is the common structure of characteristic which all living things have?

A) Having true nucleus

B) Having same number of chromosomes

C) Having cell wall

D) Having chloroplast

E) Having cytoplasm

3. Which of the following studies the animals?

A) Histology B) Zoology

C) Physiology D) Botany

E. Embryology

III. INSTRUMENTS USED IN THE BIOLOGY LABORATORY

A. MICROSCOPE

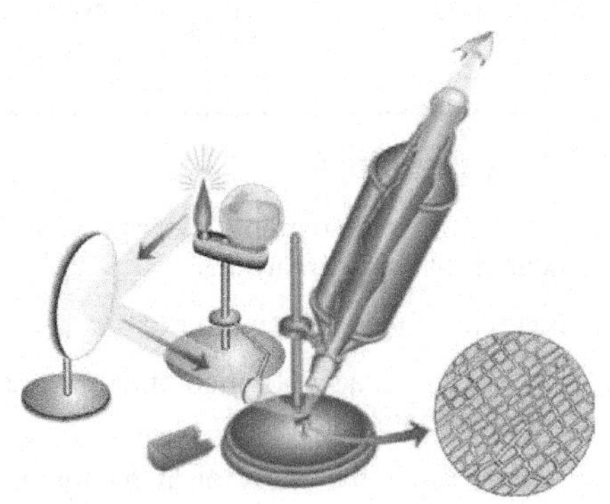

Figure 3.1.: *Robert Hooke's microscope and his observations of cork cells.*

With the naked eye we can't distinguish objects smaller than 100 μ (0.1 mm) in size. In laboratories, special tools are used to magnify such very small objects. Those tools are microscopes. Simply, microscopes allow observation of very small objects.

The first microscope was invented by Anton van Leeuwenhoek, in the early 1600s. Robert Hooke had developed the ancestor of today's microscopes by 1665. Today, many different types of microscopes, such as the phase contrast microscope, fluorescence microscope, and electron microscope, are used to bettter identify objects. However, the most commonly used microscopes are compound (light) microscopes.

1. Compound (Light) Microscope

a. Care of the microscope

- First of all, it must be known that the microscope is a delicate instrument and requires proper care. That's why the following guidelines should be observed.
- Place it on a smooth surface (e.g. a desk), away from the edge.
- If the microscope has a lamp, keep its wire out of the way.
- The most expensive parts of a microscope are the lenses. They must be kept clean. Only lens paper can be used to clean them, as needed.
- Use coverslips when studying temporary slides made with water.
- Do not smear the objectives. Be careful when using the oil immersion objective.

b. Using the microscope

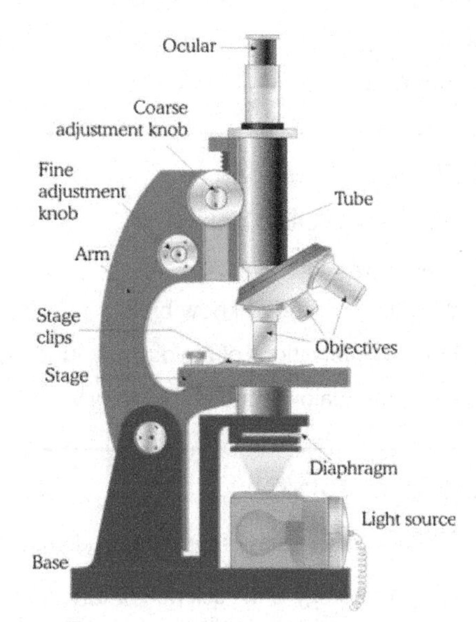

Figure 3.2.: *The compound (light) microscope*

- Hold the microscope with both hands and set it on a smooth place (table)
- Switch on the lamp (if there is one), or turn the mirror towards the light source.
- Rotate the low power objective into place.
- Some materials are best viewed in dim light, others in bright light. The light intensity can be regulated by using the diaphragm.
- Put the material to be examined on the stage under the lenses (objectives).
- Look through the eye piece.

- By using the coarse adjustment, try to focus the image.
- With the fine adjustment, sharpen the image.
- Use more powerful lenses if needed.

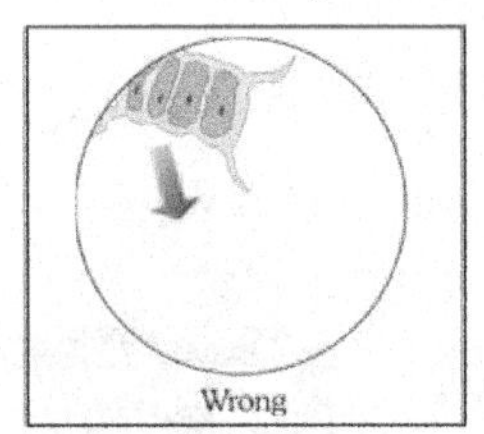

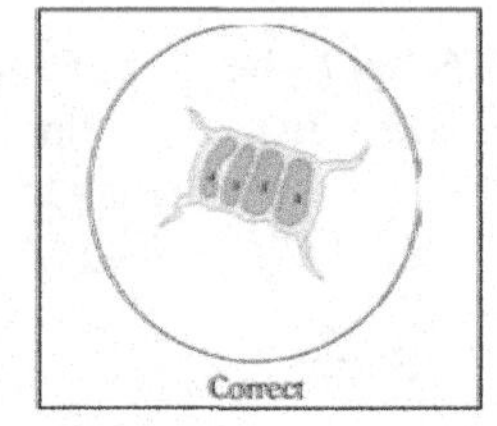

Figure 3.3.: Working with objects under the microscope

c. Magnification

The total magnification of an object is the power of the eyepiece lens multiplied by the power of the objective lens.

For example:

eye piece	objective	magnification
10x	4x (low power)	40x
10x	10x (low power)	100x
10x	40x (high power)	400x
10x	100x (oil immersion)	1000x

Table 3.1.: Total magnification of a light microscope.

B. SCIENTIFIC MEASUREMENT

In scientific research, data is gathered. In biology laboratories, mathematical data is also used. To be understood easily and fully everywhere in the world, there must be one system used by scientists. That's why, in such research, the IS (International System of Measurement) standards are used. The metric system is commonly used, in which the meter is the unit of length, the gram is the unit of mass, and the second is the unit of time.

prefixes	values	examples
kilo	1,000	a kilogram is 1,000 grams
centi	0.01	a centimeter is 0.01 meter
milli	0.001	a millimeter is 0.001 meter
micro (µ)	one millionth	a µm is 0.000001 of a meter
nano (n)	one billionth	nanogram is 10^{-9} of a gram

Table 3.2.: Common prefixes used in scientific measurements.

Among these, some that are used especially in the biology laboratory are as follows:

- **Length:** The unit of length is the meter. Other units are arranged in relation to the meter. For instance, 1 m=100 cm; centi- means 1/100, so a centimeter is 1/100 meter. That means there are 100 centimeter in 1 meter.

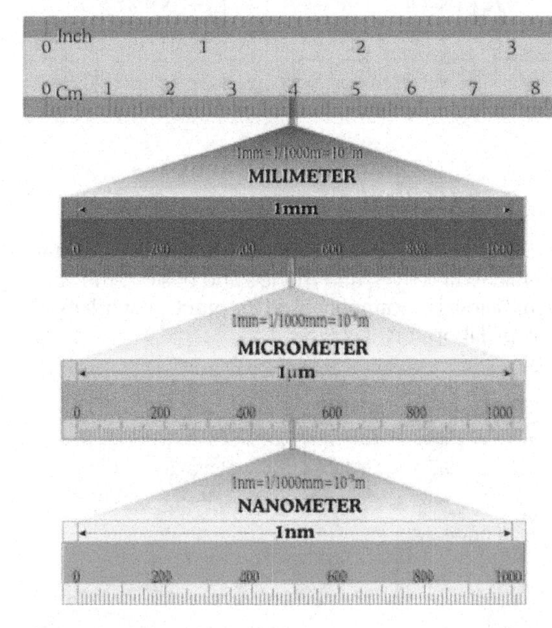

Figure 3.4.: Conversions of metric units.

- **Mass:** Mass is the amount of material in an object. The units of mass are the gram and kilogram. 1 kg=1,000 g=1,000,000 mg.
- **Volume:** Volume is the amount of space an object occupies. Its unit is the liter.
- **Time:** The unit of time is the second. In a minute there are 60 seconds, and one hour is 60 minutes.
- **Temperature:** To measure temperature, there are three familiar scales: celsius (°C), Fahrenheit (°F) and Kelvin (K). The unit of temperature commonly used all over the world is the degree celsius (°C).

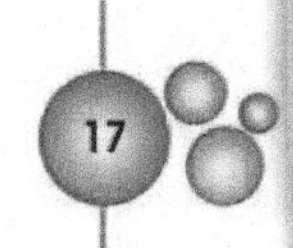

Temperature Scales	Abbreviation	Temperature Conversion
Celsius	°C	$°C=\frac{(°F-32)\times 5}{9}$
Fahrenheit	°F	$°F=\frac{°C\times 9}{5}+32$
Kelvin	K	$K=°C+273$

Table 3.3.: Temperature scales and conversions

C. GLASSWARE

Besides the microscope, dissection kits, certain chemicals and dyes, and models and posters, the glassware below is commonly used for practical work in the biology laboratory.

Figure 3.5: Some typical glassware frequently used in the biology laboratory

D. REAGENTS (INDICATORS)

In biology laboratories, special solutions and dyes are used as indicators, to identify various substances. The most commonly used indicators are:

Matter to be identified	Reagents (Indicator)	Reaction
Glucose	Benedict Solution or Fehling	Color changes to reddish brown
Starch	Iodine Solution	Dark Blue Color
Cellulose	$ZnCl_2$ Solution with Iodine	Light Blue - Green Color
Glycogen	Iodine Solution	Reddish brown
Proteins	Biuret	Purple
Proteins	Nitric Acid	Yellow
Lipids	Ether + Paper	Transparent Smear
Lipids	Sudan III	Red
Acids	Litmus Paper	Red
Bases	Litmus Paper	Blue
Acids	Congo Red	Blue
Bases	Congo Red	Red
Acids	Phenol Red	Yellow
$CO_2 + H_2O$ (Soda Water)	Phenol Red	Yellow
Breath (CO_2) (Exhalation)	Phenol Red	Yellow
$CaCO_3$ (Solution)	Lime water	White Precipitate

Table 3.4.: Some examples of indicators

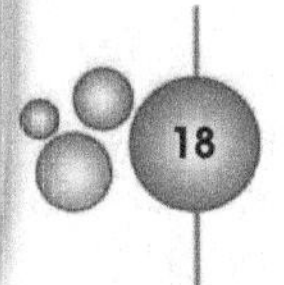

SELF CHECK

INSTRUMENTS USED IN BIOLOGY LABORATORY

A. Key Terms

Microscope	Celsius
Fahrenheit	Magnification
Kilogram	Second
Metric unit	Nanometer
Liter	

B. Review Questions

1. Briefly explain how to use the light microscope.
2. What is the function of the coarse adjustment knob on the microscope?
3. Who invented the first microscope?
4. Why are microscopes used in the laboratory?
5. What does IS (International System of Measurement) mean? Explain with an example.
6. List five kinds of glassware used in the biology lab.

C. True or False

1. ☐ 20 °C = 68 °F
2. ☐ 1 g = 1,000 mg
3. ☐ 1 km = 100 m
4. ☐ When you look at an object with the microscope under the 40X eyepiece and 40X objective, the total magnification is 4000X.
5. ☐ The unit of volume is the liter.

D. Matching

a. °C	( ) Fahrenheit
b. Fine adjustment	( ) Magnifying lenses of the microscope.
c. 1 mm	( ) 0.001 m
d. Objectives	( ) Celsius
e. °F	( ) Sharpens the image
f. Iodine solution	( ) Indicator of starch

E. Fill in the blank

1. One gram equals 1000 milligrams, and one meter equals ....
2. In standard metric units, the unit of length is the .... (m), the unit of mass is the .... (....) and the unit of time is the .... (....).
3. One liter is equal to .... milliliters.
4. The prefixes kilo-, centi- and milli- have values of 1,000, ... and ....

F. Multiple choice

1. Moves the body tube a whole lot

A) diaphragm B) stage
C) coarse adjustment D) fine adjustment
E) spring clip

2. What holds the slide on the stage

A) diaphragm B) stage
C) coarse adjustment D) fine adjustment
E) spring clip

3. If the eyepiece of the microscope magnifies 10 times and the low power objective magnifies 45 times, the total magnification under low power is

A) 45 times B) 55 times
C) 450 times D) 400 times
E) 600 times

IV. CELLULAR ORGANIZATION

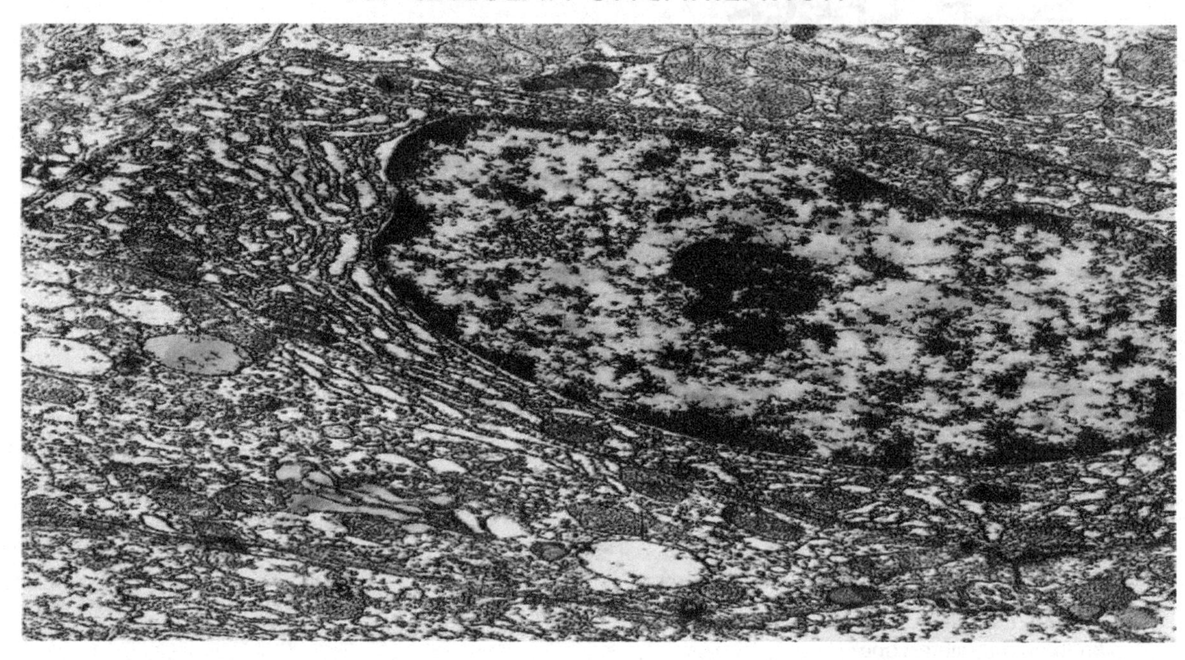

A. CELLULAR ORGANIZATION

Cells are the fundamental units of life. They are the structural and functional units of living organisms. This explanation states that:

- All living organisms are made of cells.
- Each cell originates from another cell.
- All activities going on in the bodies of living organisms take place within cells.

According to cell number, organisms are classified as unicellular and multicellular. Unicellular organisms are made of only one cell. One independent cell can do everything needed to survive. Members of the kingdoms monera (bacteria and cyanobacteria/blue-green algae) and protist are unicellular. However, more complex organisms are made of very many cells and are called multicelluar organisms.

In multicellular organisms, all cells depend on one another. There is perfect organization and division of labor between the parts of the body.

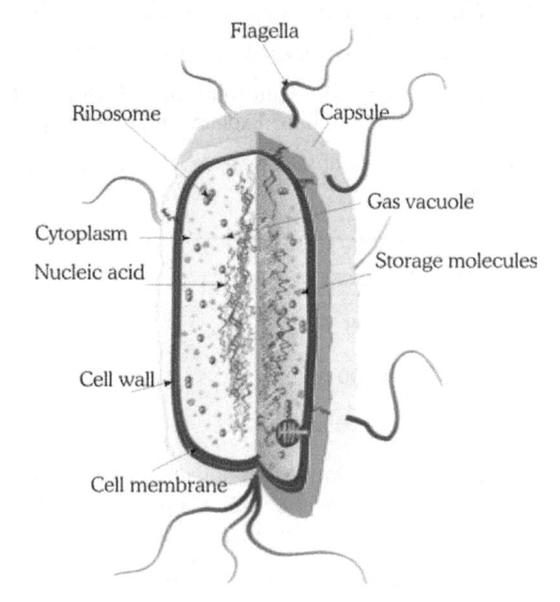

Figure 4.1.: A typical prokaryotic cell, a bacillum.

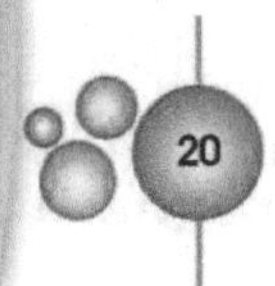

Besides the cell number, cell complexity (structure) is another criteria used to classify organisms.

According to their structure, cells may be prokaryotic or eukaryotic. Prokaryotic cells (from the Greek pro-, meaning "before", and ***karyon***, meaning "kernel" or "nucleus"), the modern bacteria and cyanobacteria, were the first life forms 3.5 billion years ago. They are still abundant and diverse, but very limited in size and metabolic complexity. While most prokaryotes are 1-10 μm in diameter, also included are the smallest known cells, the mycoplasmas, with diameters between 0.1-1.0 μm. Prokaryotic cells have no nuclear membrane or membrane-bound organelles. The DNA of prokaryotes is concentrated in a region of the cytoplasm called the nucleoid. Members of kingdom monera are prokaryotes.

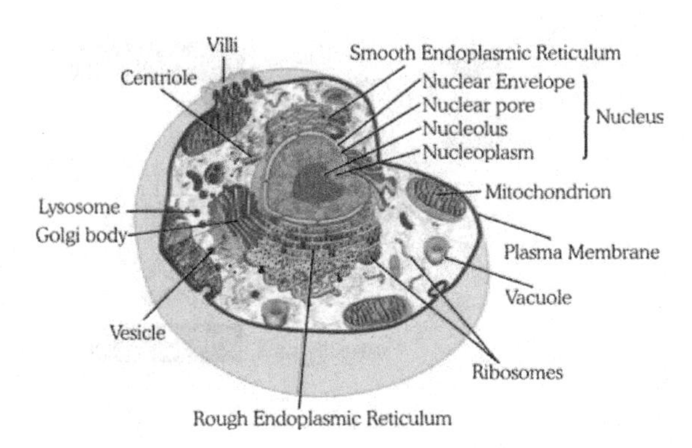

Figure 4.2.: Typical animal cell structure.

Eukaryotic cells (from the Greek eu-, meaning "true", hence, having a true nucleus) first appeared roughly 1.5 billion years ago and now constitute all organisms other than bacteria and cyanobacteria. With diameters from 10-100 μm they are, on average, 10 times larger than prokaryotic cells. As mentioned before, eukaryotic cells have organelles with specific functions, e.g., the nucleus synthesizes RNA and DNA, mitochondria and chloroplasts transform energy into useable forms, lysosomes digest macromolecules, and ribosomes bound in the endoplasmic reticulum and free ribosomes suspended in the cytosol synthesize proteins. Because they have membrane bound organelles with specialized functions, eukaryotic cells are complex. These properties allow eukaryotic cells to use environmental resources not available to prokaryotic cells.

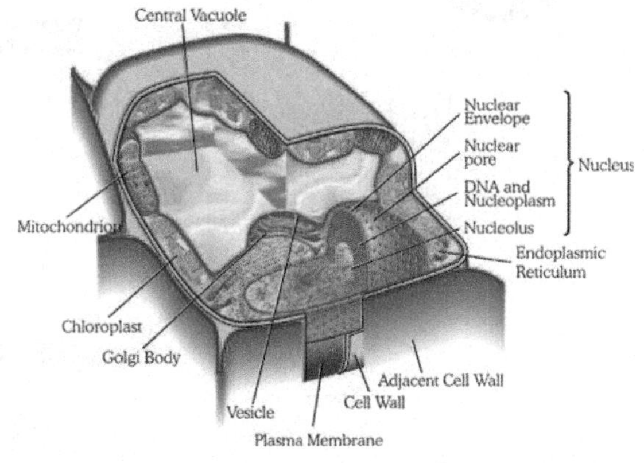

Figure 4.3.: Typical plant cell structure.

The best known eukaryotes are plants and animals. According to cell structure, they are categorized in the same group, as eukaryotes, but these two well-known groups of living organisms have certain differences in the structure and function of their cells. These differences can be summarized simply as follows:

Animal Cells	Plant Cells
No cell wall, only a cell memrane.	Cell wall in addition to cell membrane.
Numerous, small vacuoles	Fewer but larger vacuoles
Have centrosomes: responsible for spindle fiber production during cell division.	Spindle fibers are formed by some special proteins found in the cytoplasm.
Don't have plastids.	Plastids (Leucoplasts, Chloroplasts, and Chromoplasts) are found
Heterotrophs (consumers)	Autotrophs (producers)
Generally oval shaped	Generally rectangular shaped

Table 4.1.: Comparision of animal and plant cells

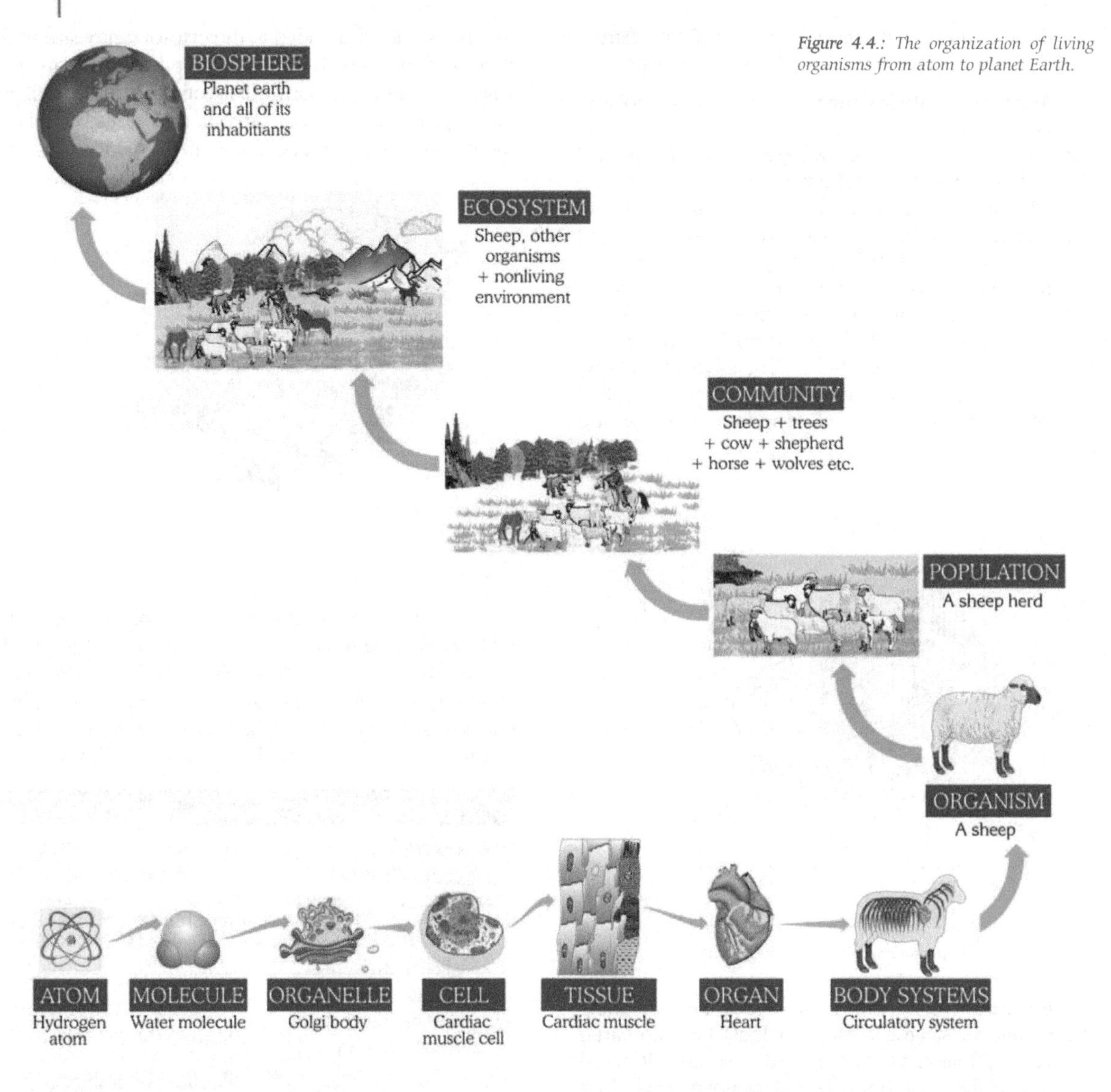

Figure 4.4.: The organization of living organisms from atom to planet Earth.

B. MULTICELLULAR ORGANIZATION

Among eukaryotes, many are multicellular, or **"made of many cells"**. In multicellular organisms, a group of cells with some certain similar structural and functional properties form TISSUE. Tissue is made of some cells and intercellular fluid between them. There are approximately two hundred different cells and four main tissues in animals. Main animal tissues are:

- Epithelial tissue
- Connective tissue
- Muscular tissue
- Nervous tissue

Blood is a special type of connective tissue.

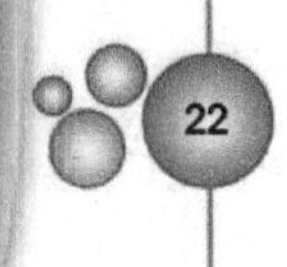

Tissues that work together for a specific function form an ORGAN.

The heart is an organ.

Organs come together to form ORGAN SYSTEMS.

Heart, blood, and blood vessels all together form the circulatory system, which is responsible for the distribution of substances around the body.

All the organ systems together form an ORGANISM.

In multicellular organisms, there is a division of labor, meaning each part of the body has unique function and all cells work in dependence on each other.

C. LIFE AND LIVING ORGANISMS

When the definition of biology and information about its sub-branches are given, it is mentioned that biology is the science of living organisms and the interactions between them. However, the properties of living things and the differences between them haven't been mentioned yet. Clarifying these things will help us to understand almost everything about the fundamentals of biology and living organisms much better.

Organisms that have the properties of life are called **"Living Organisms"**. Some non-living things, like iron and stones, have some properties of life, such as atomic energy and mobility (at least at the atomic level). Viruses, that are crystals in isolation, show many life activities when they are with real living organisms.

All these suggest that very clear differences between living and non-living forms can't be demonstrated. However, beside this fact there are some certain characteristics of life, and members of the monera, protist, fungus, plant and animal kingdoms, that have these properties, are called living organisms. Even though different types of living organisms may have some unique structural differences, the characteristics that wil be mentioned below are common in all.

1. Chemical Makeup

Structures of all living organisms are made of similar chemical substances. Aproximately 65% water, 2% carbohydrates, 15% proteins, 7% lipids, some minerals (Ca, Mg, P, Fe, N, etc.), and nucleic acids (DNA and RNA).

These chemical substances are used as sources of energy, structural elements or regulatory substances within the bodies of living organisms. For continuity of life processes these substances have vital importance.

2. Organization

Cell → Tissues → Organs → Body Systems → Organism

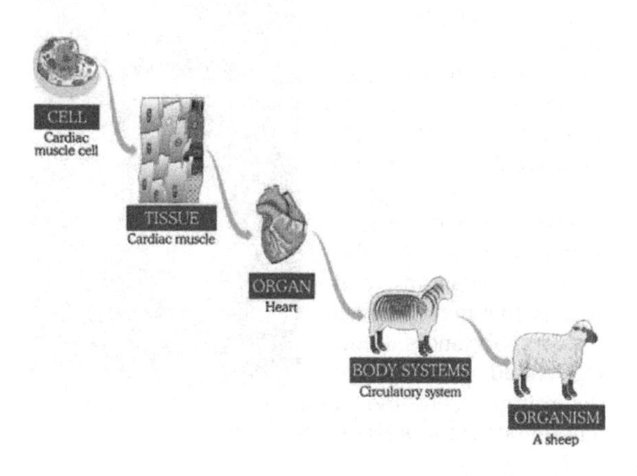

Figure 4.5.: Organization of living things cell to organisms.

In the body of an organism each part provides a specific function. As a result, substances and energy are used economicaly and the body works in harmony. Organization of unicellular organisms is achieved by performance of different functions by different parts of the single cell. In multicellular organisms, structural and functional units are cells. Cells form tissues, tissues form organs, organs form systems, and systems form organisms. Each part of the body of an organism posseses a unique function.

3. Energy Production (Cellular Respiration)

All life activities, such as reproduction and growth, require energy. The main source of energy used by all living things is the sun. By means of photosynthesis radiant energy is converted into chemical bond energy in the form of organic substances. Then this energy is converted into useable energy in the process called cellular respiration.

4. Nutrition

All living things have to feed themselves to survive. As mentioned before, energy is needed for continuity of life processes, and this energy is supplied from ingested organic substances. Some organisms can produce food for themselves and others. Such organisms are called autotrophs. Autotrophs produce organic substances either as a result of photosynthesis (green plants) or chemosynthesis (some bacteria). Another group of organisms, consumers, that don't produce organic substances (food) are called heterotrophs. Heterotrophs consume food produced by other organisms, either autotrophs or other heterotrophs.

5. Growth

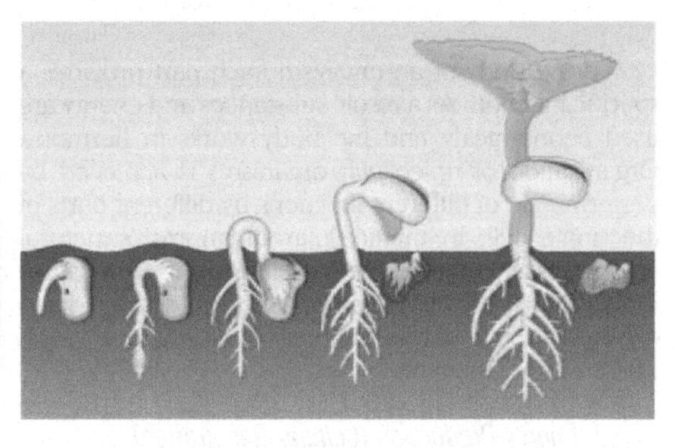

Growth is a life process in which the cells of living organisms increase in size, number and mass. Growth takes place by means of ingestion of nutrients and cellular respiration. Use of organic substances within the cells which gives rise to energy production is called metabolism. Metabolism is simply all the biochemical processes that take place in the cells of the body. These biochemical processes are categorized as two main types: anabolic and catabolic reactions. Anabolic reactions are synthesis reactions. Ingested food is digested by the organs of the digestive system, then digested food particles are absorbed into the blood and transported to body cells, where they are used. In the cells, these monomers are synthesized and become various parts of the cell. On the other hand, in catabolic reactions, ingested food is catabolized within the cell, providing energy and other vital substances needed by the cell.

If the rate of anabolic reactions is greater, the organism grows. If anabolic and catabolic reactions are the same, there is no growth. If the rate of catabolism is higher, the organism is aging. However, growth is a unique process for each organism and is limited by the genetic properties of the organism. The limits to growth in animals are more clear than those in plants.

6. Movement

Related to their ability to move, living organisms have different strategies for movement. These are found at the cellular, organ, system, and organism levels. Cytoplasmic movement in a cell, muscular movement of animals, and turning of plants towards sun light can be given as examples of different types of movement in organisms.

In animals and humans, movement of the organism is caused by the muscular and skeletal systems.

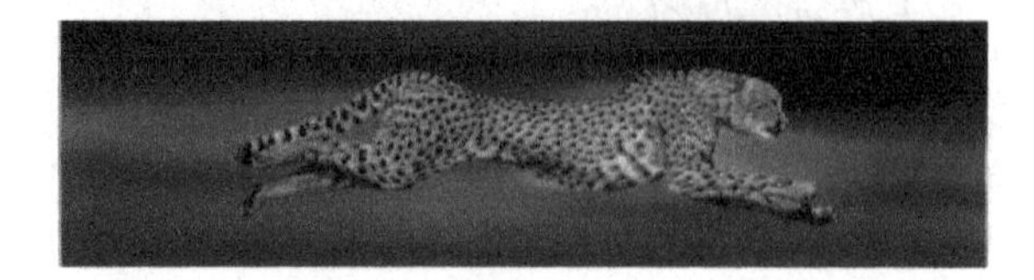

7. Irritability "Sensitivity"

Organisms respond to the physical and chemical stimuli received from their environment as well as the internal parts of the body.

Leaves turning towards light, and changes in the diameter of the pupil when light intensity is changed are some examples that illustrate this complex ability of organisms.

Sensitivity of organisms is controlled by the nervous system, endocrine system and sense organs.

8. Reproduction

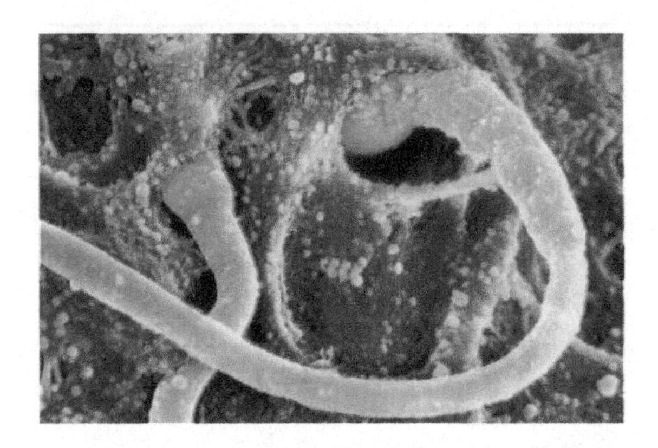

All living things must be able to reproduce. Through reproduction, the species continues its existence. All hereditary traits that determine a species and make it fit for a particular environment are transmitted to the offspring by means of reproduction.

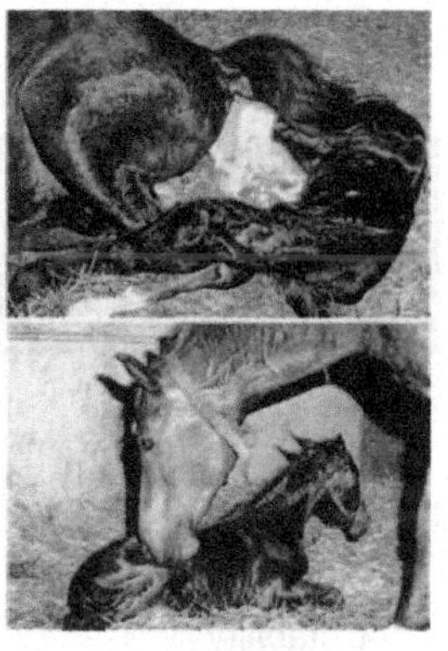

Organisms reproduce sexually or asexually. Both are based on cell division. There are two types of cell division, mitosis and meiosis. In both sexual and asexual reproduction, mitosis takes place, but meiosis is cell division unique to sexual reproduction. As a result of mitosis, two genetically identical daughter cells are produced from one parental cell. In meioisis, one reproductive mother cell gives rise to four gametes with some genetic differences.

9. Homeostasis (Internal Balance)

Even though the external environment is continously changing, the body tries to keep itself in a steady state. This balance is called homeostasis. All parts of the body have responsibilities for keeping the body in homesotasis, but overall control is the responsibility of the regulatory systems (nervous and endocrine systems).

10. Death

All organisms with the characteristics mentioned above are born, grow and die. Death is the end of life processes. Each organism has a certain life span. Under normal conditions, organisms complete their life span and die.

SELF CHECK

CELLULAR ORGANIZATION

A. Key Terms

Cell	Multicellular
Nutrition	Animal Cell
Irritability	Unicellular
Growth	Plant Cell
Organism	Homeostasis

B. Review Questions

1. What does cell mean?
2. Compare Animal cell and Plant cell.
3. Compare Prokaryotic and Eukaryotic cell structure.
4. How can we distinquish living things and non-living objects?
5. Show the organization of living things from cell to biosphere.

C. True or False

1. ☐ A group of cells with some certain structural and functional similarities form a tissue.
2. ☐ Metabolism is the process in which reproductive cells are produced.
3. ☐ Each cell originates from another cell.
4. ☐ In multicellular organisms, there is a perfect organization and division of labor between the parts of the body.

D. Matching

a. Population	( ) Planet Earth & all of its inhabitants.
b. Homeostasis	( ) Formed by same kind of cells.
c. Heterotroph	( ) Internal body balance
d. Cellular Respiration	( ) Contains only one kind of organism
e. Tissue	( ) Energy production process in living organisms.
f. Biosphere	( ) Consumer.

E. Fill in the blank

1. Irritation of organisms is controlled by ...., ... and sense organs.
2. Living things reproduce .... or ....
3. All organisms .... , ... , and die in their life span.
4. In addition to having a cell membrane, plant cells are surrounded by a ....

F. Multiple choice

1. In prokaryotic cells there is no __________.

A) cell membrane B) nucleus

C) cytoplasm D) organic molecules

E) DNA

2. Which of the following is not property of eucaryotic cells

A) They have cell membrane

B) They have no nucleus

C) They can make colony

D) They have membranous organelles

E) They have ribosomes

V. BODY STRUCTURE IN LIVING ORGANISMS

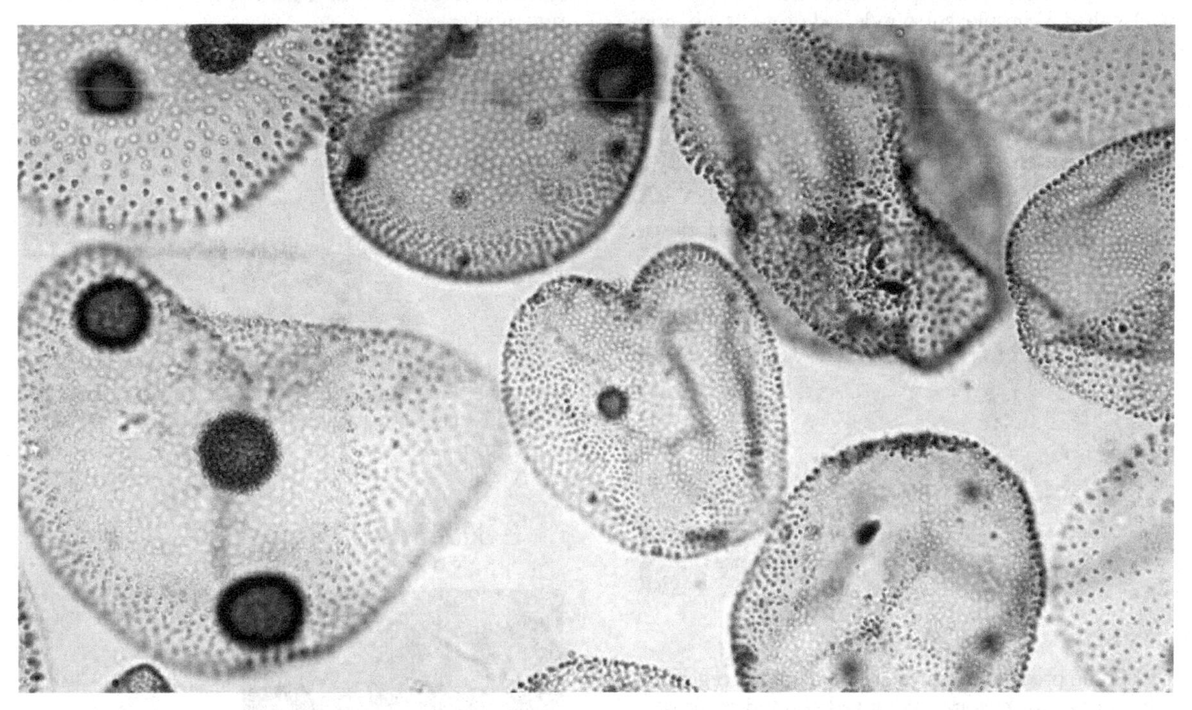

1. BODY STRUCTURE IN MICROORGANISMS

Scientists categorize microorganisms into three major groups: viruses, monerans and protists.

Viruses don't have cellular organization, so they can not perform all the metabolic activities of a true cell. Viruses are parasites and always need other organisms, "host organisms", to continue their life. In unfavorable conditions, most viruses crystallize. When they are crystallized, viruses don't posses the features of living organisms.

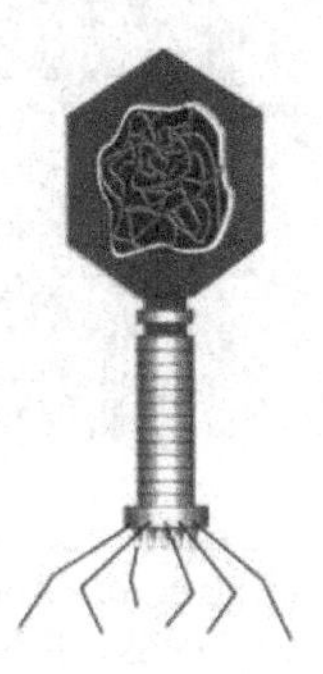

Figure 5.1.: *A bacteriophage virus*

The diameter of viruses ranges from 10 to 300 nm (nanometer). Viruses can be seen only with the scanning electron microscope.

A virus is made up of two main components, a rounds the core consisting of either DNA or RNA. Viruses lack cytoplasmic content and organelles.

Some viruses have a capsule around them to protect them from undesirable conditions like high temperature and acidity.

Viruses are all pathogenic on plants, animals or humans.

Blue-green algae and bacteria form the entire monera kingdom. A typical bacteria reflects all the common characteristics of monerans.

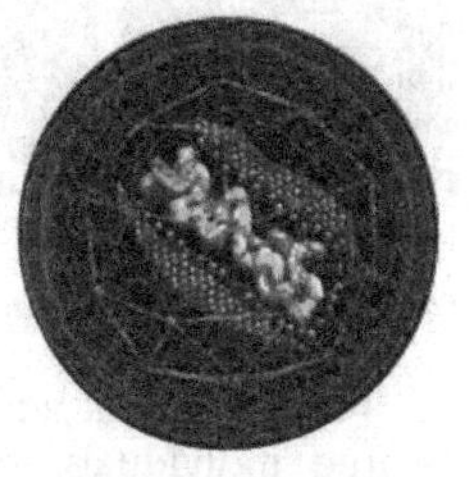

Figure 5.2.: *HIV, the virus that causes AIDS*

Bacteria are prokaryotic, unicellular organisms. The cell structure is clearly different in having no true nucleus nor membrane surrounded organelles. Bacteria contain both DNA and RNA as genetic material, dispersed in the cell cytoplasm. Prokaryotic ribosomes, unique to bacteria, are essential in producing the proteins necessary for bacteria. The rigid cell wall, called a capsule, surrounds and protects the monerans from undesirable conditions. Many monerans are motile, with one or more flagella. Some can produce their own food via photosynthesis or chemosynthesis. The photosynthetic members possess chlorophyll, to convert solar energy into the chemical bond energy in food, and look green.

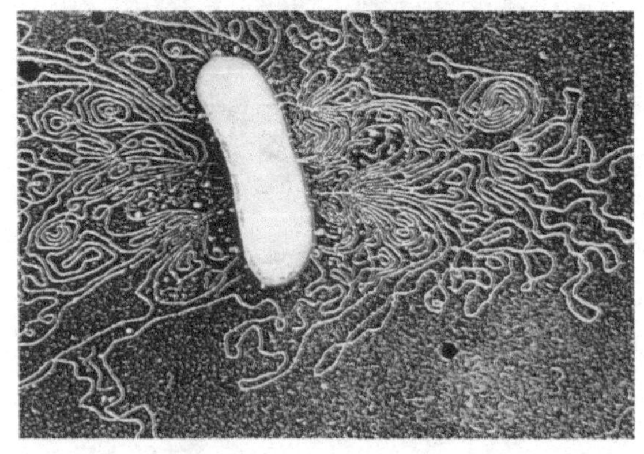

Figure 5.3.: A bacterium

Some protists are unicellular, eukaryotic organisms with a true nucleus and membrane-bound organelles. Some members, such as Euglena, have chloroplasts to produce their own food by photosynthesis. The rest are heterotrophs and feed on other organisms. Protists have flagella, cilia or pseudopodia for movement and food collection.

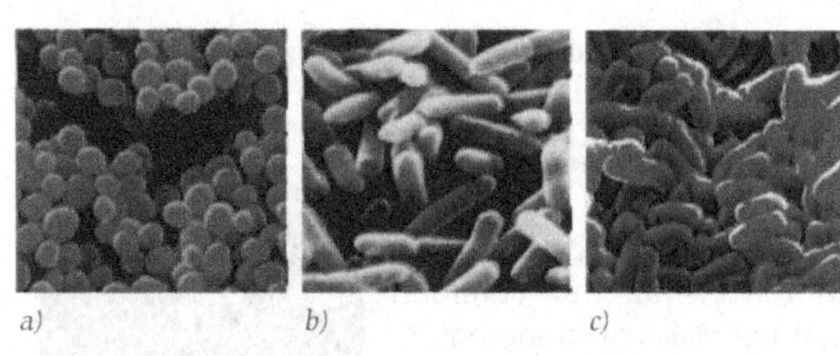

a) b) c)

Figure 5.4.: a) Cocci, b) Bacilli, c) Mycobacterium tuberculosis

Protists inhabit aquatic habitats. They usually live as free individuals, but protists also form simple colonies from a group of collaborating, similar cells. However, some protists can form complex colonies in which there is an association between cells to form a multicellular organization. In complex colonies, there is a division of labor, e.g., the outer-most cells are responsible for locomotion and getting food, the others for digestion.

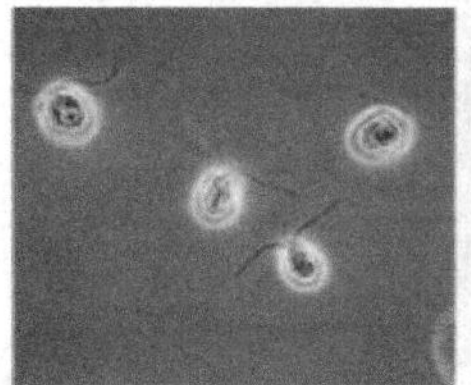

Figure 5.5.: Free living chlamydomonas.

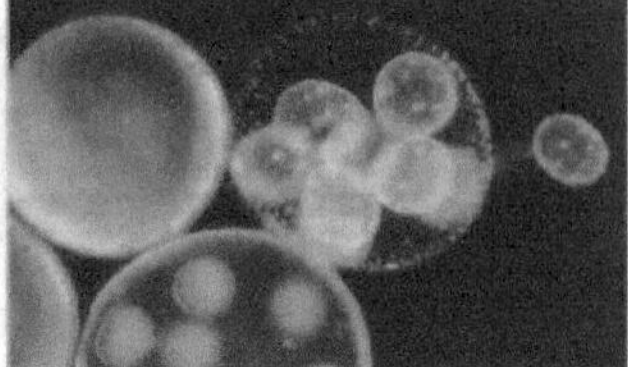

Figure 5.6.: The volvox colonies.

For example, chlamydomonas may exist as free-living individuals. When they form the volvox colony, the individual cells are embedded within the colony, separated from each other by a gelatin layer surrounding every cell.

2. BODY STRUCTURE IN FUNGI

Figure 5.7.: Mycelia of zygomycota (bread mold), with many sporangia.

Except unicellular yeast, all fungi have eukaryotic cell structure and multicellular organization. Except for yeast cells, all fungi have hyphal structures: colorless, slender, branched, thread-like tubes. In multicellular fungi, the total hyphal structure forms a mass, like a body, called a mycelium. In some fungi, the hyphae are continuous threads of cytoplasm with many nuclei. Most fungi have a cell wall around their cells and plastids, like plants, but they do not have chloroplasts. All fungi are heterotrophs and exclusively saprophytic.

Certain types are parasites on plants and animals. Most live in aerobic conditions, but there are also some anaerobic forms.

Figure 5.8.: Athlete's foot fungus.

Lichens are the symbiosis (common life form) of fungi and blue-green algae. In this common life form, both benefit from each other. Fungi collect water, and blue-green algae contain chlorophyll for photosynthesis. They produce and share food.

Most fungi reproduce with spores produced in sporangia located at the tip of mature hyphae.

Figure 5.9.: A mushroom is a member of basidiomycota.

Spores can form whole organisms when conditions are favorable.

If conditions are not good, spores stay in dormancy, which can persist until favorable conditions arise. The dormancy state may last many years. Some fungi are capable of a form of sexual reproduction in which there are no male and female traits. Some hyphae form (+) traits, others forms (-) traits. The (+) and (-) traits fuse to form a zygospore, which grows into a young organism.

3. PLANT BODY STRUCTURES

Plants are multicellular, eukaryotic organisms that can synthesize their own food by trapping solar energy and storing it in the form of chemical bonds. Over 270,000 species have been identified. Remarkably, about 90% are flowering plants with seeds enclosed in fruit. Plants inhabit the entire earth's surface in different locations and climates, like cacti in deserts, rhubarb, onion and asparagus in temperate climates, and palms and orchids in tropical zones.

Regardless of size, from the smallest flowering plants (*Wolffia microscopica*) to the tallest (redwood and sequoia trees), all have the same basic body plan.

Most plants are either herbaceous or woody. Some, known as annuals, complete their life cycle in one year and are herbaceous (geranium, corn, bean, etc.). Other herbaceous plants (carrot, cabbage, beet etc.) are biennials that grow, reproduce and die in two years. Perennials live many years, have woody bodies, and are larger in size.

A true plant has three main body parts:

- Root
- Leaf
- Stem

A. ROOT

Roots absorb water and minerals from the soil and anchor the plant in the ground. In some plants like carrot and beet, it also stores food. The typical root structure is seen in the figure. The tip of the root is rounded and covered with a protective structure called a root cap. Just beneath the protective root cap, are the apical meristematic cells. These divide to form new cells, resulting in the longitudinal growth of roots. The root hairs, tiny projections, grow out from the mature parts of the root. These hairs increase the surface area in direct contact with soil through which water and minerals may be absorbed.

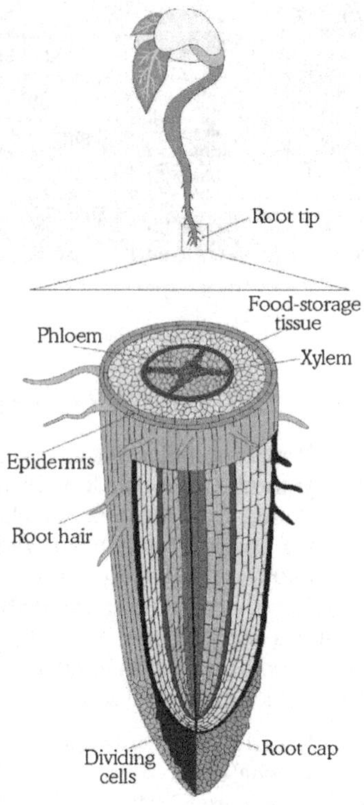

Figure 5.10.: Typical root structure

Vascular (transport) tissue is located at the core of the root. Water and minerals are absorbed into the xylem vessels, then transported uni-directionally upward to the stem and leaves of the plant.

The phloem vessels transport the food produced in the leaves to the other parts of plant in a bi-directional manner: upward and downward.

A cambium layer that produces new xylem and phloem vessels is located between the xylem and phloem.

Production of new xylem and phloem results in growth of the plant's diameter.

B. LEAVES

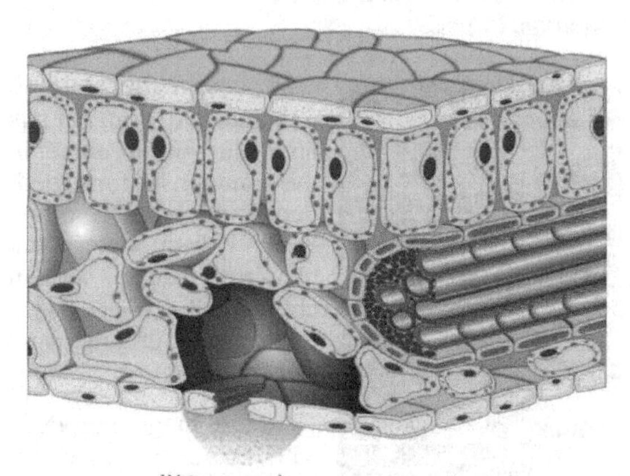

Leaves are the most numerous parts of many plants. Leaves vary in size and shape: spiny leaves in pine, reduced leaves in cacti, broad leaves in tropical plants, ribbon-like leaves in corn, aquatic leaves in water lily. Leaves play an important role in capturing solar energy and converting it into the chemical energy in food.

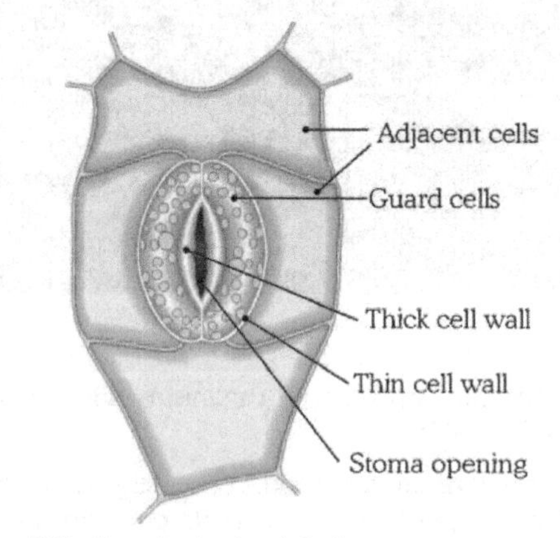

Figure 5.11.: Stomata structure in leaf.

When a cross section of a leaf is examined under the microscope, the outermost layer is a waxy, waterproof cuticle layer. Inside the waxy layer, surface cells form a single cell layer. Next come the upper leaf cells containing chloroplasts for photosynthesis. At the center of the leaf is found vascular tissue composed of xylem and phloem vessels. The cells on the underside of the leaf have many spaces to temporarily store CO_2 and O_2. The underside of the leaf is covered by a single layer of cells with tiny pores, called stomata, for gas exchange.

Water is absorbed by the roots and travels through the xylem. Carbon dioxide enters through the stomata. During photosynthesis, sugar and O_2 are produced, and CO_2 and water are consumed.

C. STEM

The stem connects the roots and leaves. Stems can be either herbaceous or woody. Herbaceous stems are soft and can undergo photosynthesis. Bean, wheat, tomato, dahlia, and pepper plants have herbaceous stems.

On the other hand, woody stems are hard and non-photosynthetic. Maple, pine, oak trees and roses have woody stems.

Both herbaceous and woody stems are composed

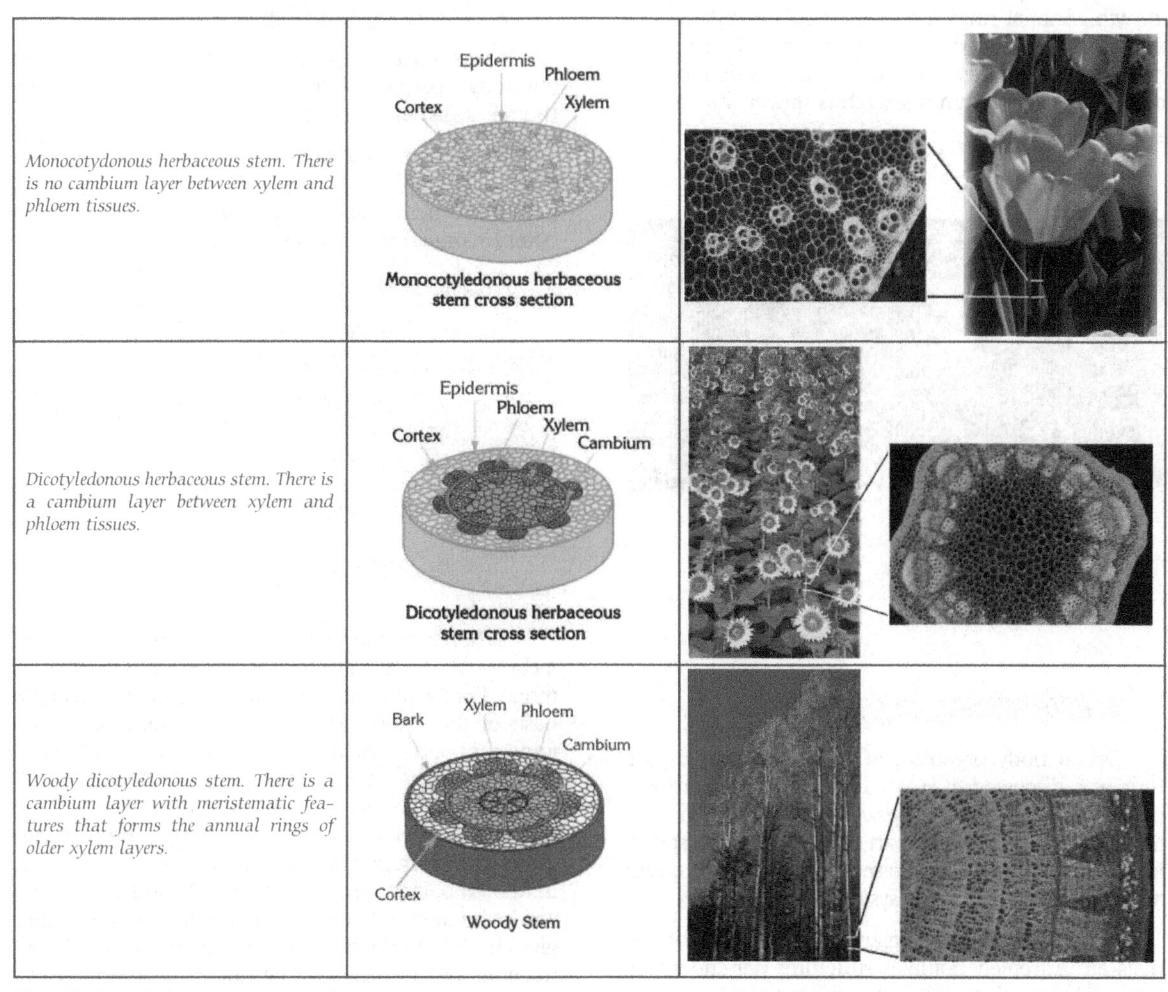

Monocotydonous herbaceous stem. There is no cambium layer between xylem and phloem tissues.
Epidermis
Phloem
Cortex
Xylem
Monocotyledonous herbaceous stem cross section
Dicotyledonous herbaceous stem. There is a cambium layer between xylem and phloem tissues.
Epidermis
Phloem
Xylem
Cortex
Cambium
Dicotyledonous herbaceous stem cross section
Woody dicotyledonous stem. There is a cambium layer with meristematic features that forms the annual rings of older xylem layers.
Xylem
Phloem
Bark
Cambium
Cortex
Woody Stem

Figure 5.12.: Typical stem structures in plants.

of xylem and phloem tissue, with some other supportive tissues in woody stems. The outer layer, called bark, helps to protect the inner tissues. The phloem is located beneath the bark. Inside the phloem is a layer called the cambium, whose cells divide to form new xylem and phloem vessels. Inside the cambium, the active xylem tissue is found. At the core is found the heartwood, old inactive xylem tissue. In the center of the stem there is pith.

When you look at the stump of a tree, you see many concentric circles. These are called annual rings, because each pair of them represents one year of plant growth. Annual rings are formed by the xylem. The xylem cells produced in spring form a wide, light-colored ring. In contrast, the summer ring is thinner and darker, because summer growth is slower than spring growth.

4. BODY SYSTEMS IN ANIMALS

When body organization in multicellular organisms was discussed, it was mentioned that there is a perfect division of labor between body parts in such organisms, meaning each part of the body has a specific function to do. It was also mentioned that organisms are formed of many organ systems.

Among multicelluar organisms, the bodies of animals are extremely complex structures dependent on the synchronization of the following systems:

- [illegible]
- [illegible]
- [illegible]
- [illegible]
- [illegible]
- [illegible]
- [illegible]
- [illegible]

A. [illegible]

It was already explained that there are systems in the body working in harmony, but how is this balance (homeostasis) maintained?

This section will explain that the [illegible] between organs, and regulate important functions in complex organisms, such as animals.

1. The [illegible]

Among complex organisms, [illegible]

The [illegible]

[illegible]

[illegible]

When a hormone is received by the target organ, it causes changes in its function, it responds to the hormone. For instance, the pituitary is a gland found at the base of the brain just under the hypothalamus (an important part of the nervous system). It is a very small gland. [illegible] Oxytocin affects the muscles of the uterus, the part of the female's body where babies grow, and mammary glands, where milk is produced. In human beings, a baby grows for around forty weeks inside the uterus. At the end of this period, the mother gives birth to the baby. This is caused by powerful contractions of the muscles that cover the uterus. Now,

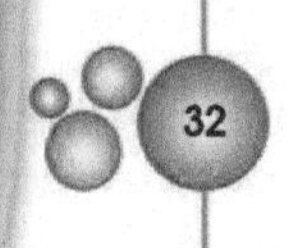

how do these muscles know that they must contract forcefully? It is the function of oxytocin to communicate to the muscles what needs to be done. At the time of birth, oxytocin is secreted by the pituitary gland. It is brought to the muscles via the blood, and the muscles contract. Indeed, this extra muscle activity causes labor pains, but this is the only way to give birth to a baby.

Another function of oxytocin is to cause contraction of the mammary glands. After birth, with the help of prolactin, another hormone of the pituitary gland, milk is produced by the mammary glands. However, the baby doesn't know how to suckle. When the baby touches its mother's nipples, the pituitary gland is stimulated, oxytocin is produced, the mammary gland contracts, and milk is squirted into the baby's mouth.

Just like the pituitary gland and oxytocin, the bodies of animals and human beings have many glands and hormones that regulate various functions.

2. Nervous System

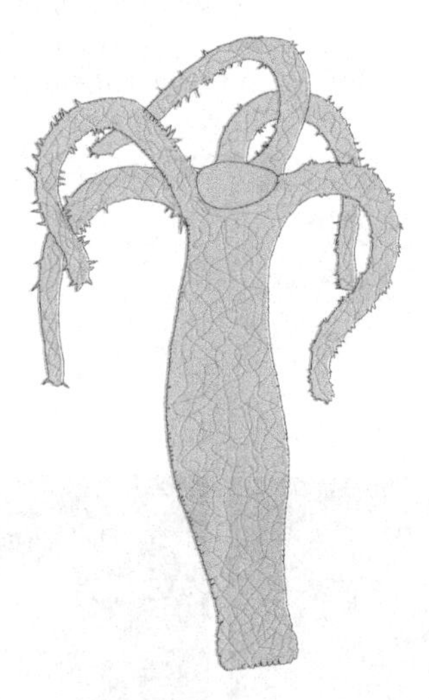

Figure 5.13.: *Organization of the nervous structures in hydra*

We have already learned that the internal balance of an organism, homeostasis, must be maintained. As one of two regulatory systems, we have learned how the endocrine system works. The other regulatory system is the nervous system.

Nervous connections to all organs of the body provide information on changes in the internal environment. The information is evaluated and

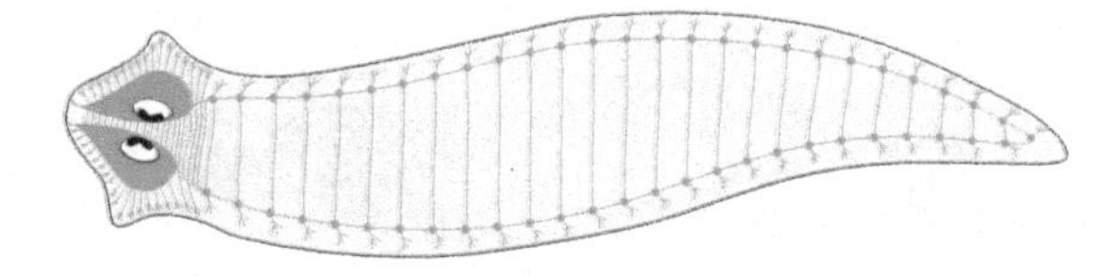

Figure 5.14.: *Planarian nervous organization*

The nervous system is made of nerve tissue. Nerve tissue includes cells and interstitial fluid. The well-known, functional

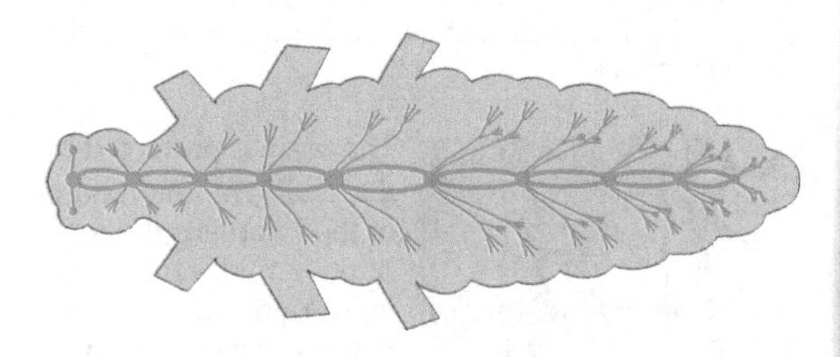

Figure 5.15.: *Nervous organization in insects.*

which are short, thin projections

the main part of a neuron

, a single, long extension from the cell body

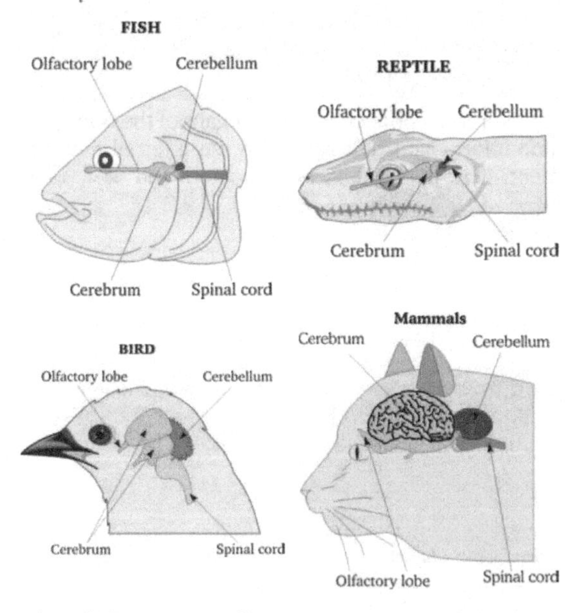

Figure 5.16.: Comparison of brain structure among vertebrates

As an example of complex neural organization, the human nervous system consists of two main divisions, the central and peripheral nervous systems. The central nervous system (CNS) consists of the brain and spinal cord. The brain is the most important part of the nervous system.

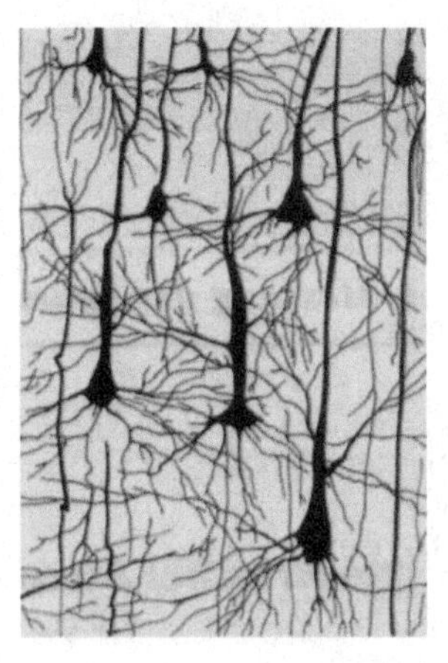

Figure-5.17: Branched nerve cells

The brain is protected by the skull. The main parts of the brain are the forebrain, midbrain, and hindbrain. Each has many unique parts with different functions. Among the parts of the brain, the most complex is the cerebrum, with five lobes, each responsible for one of the very complex activities of the body, such as conscious life activities, thinking, memory, etc.

The spinal cord, the second part of the central nervous system, is the center of reflexes. A reflex is the transmission of impulses generated by receptors, to the target, where a sudden response is generated.

The central nervous system is the main part of the nervous system, but it must be connected to all other parts of the body. The nerves that provide this connection form the peripheral nervous system.

B. SENSORY RECEPTION (The Sense Organs)

An organism can only survive if it is able to detect changes both in its environment and in its body. Physical and chemical changes in the environment stimulate receptors in the structures of special organs called sense organs. Nerves within these organs transmit impulses to the central nervous system, and the response is transmitted to the effector organs by other nerves.

The CNS is informed of stimuli in the form of light, temperature, sound, taste, touch, etc. The coordination of receptors, intermediate structures and the CNS is important for the successful transmission of an impulse. An impulse can not be generated if the receptor is non-functional.

There are three main types of receptors in animals:

- Photoreceptors
- Chemoreceptors
- Mechanoreceptors

1. Photoreceptors

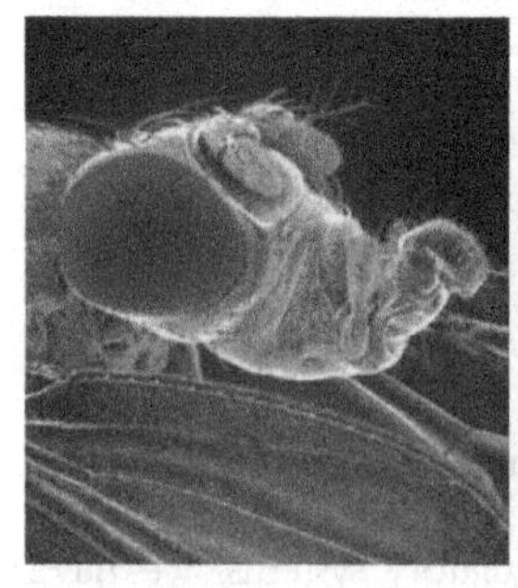

Figure 5.18.: Eye structures in Drosophila and house fly.

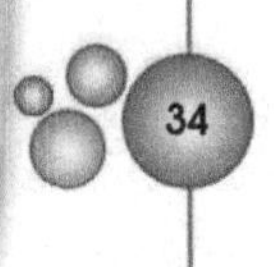

Photoreceptors are found in the eye and are light sensitive. The human eye is capable of detecting light between 4000 and 7400 Å.

Different organisms have different types of eyes for detecting light. The eyes of vertebrates are similar to each other in structure and function, resembling a simple camera. The eyeball and accessory structures are parts of the vertebrate eye. The eyeball is made of three parts:

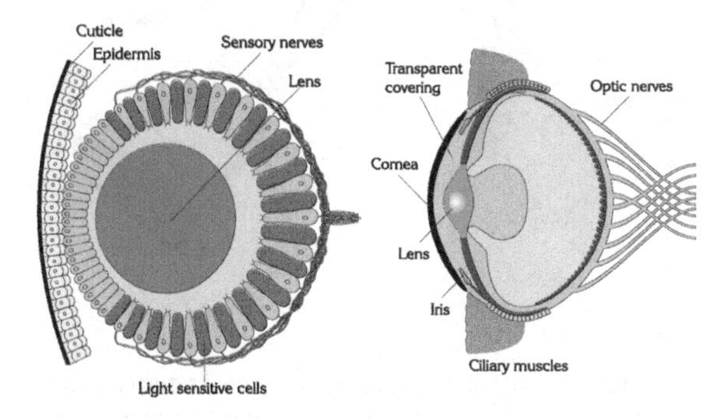

Figure 5.19.: The structures involved in sight.

- Sclera
- Choroid
- Retina

The sclera is a supportive, protective structure. The choroid is rich in blood vessels and pigments. Its main function is to supply nutrients to the eye. The retina is the innermost layer of the eyeball and is the part where photoreceptors are found. Stimuli arriving here are transmitted to the CNS, and vision is produced.

2. Chemoreceptors

Chemoreceptors are found in two of the sense organs, the nose and the tongue. Chemoreceptors are sensitive to dissolved chemical suhstances. The chemoreceptors located in the nasal cavity are stimulated by chemicals dissolved in the nasal mucosa. The chemoreceptors found here are connected to the CNS by olfactory nerves. In the CNS these chemicals are interpreted as different odors.

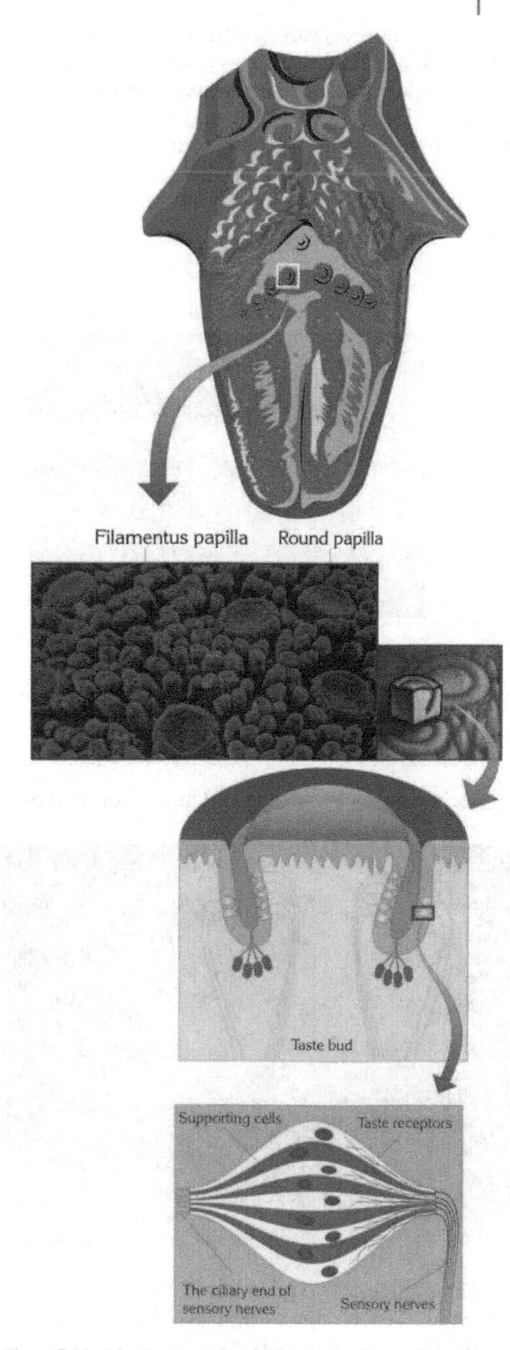

Figure 5.20.: The structures involved in tasting.

The chemoreceptors found in the tongue are located within special structures called taste buds. The human tongue contains approximately 9-10 thousand taste buds. The tongue is divided into four regions involved in the sensation of taste: bitter, sour, salty and sweet. Taste buds located in different parts of the tongue detect different tastes. These stimuli are sent to the CNS to be interpreted.

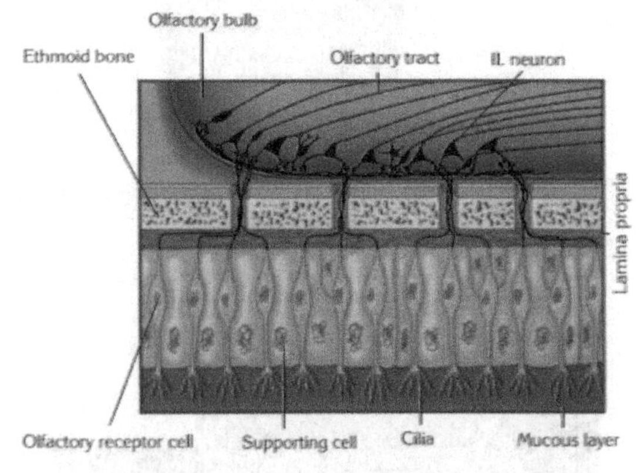

Figure 5.21.: The structures involved in smelling.

3. Mechanoreceptors

Mechanoreceptors are found in the skin and ears.

Skin is made of two main layers, the epidermis and the dermis. The receptors that provide sensation of heat, pressure, touch etc. are found in the dermis.

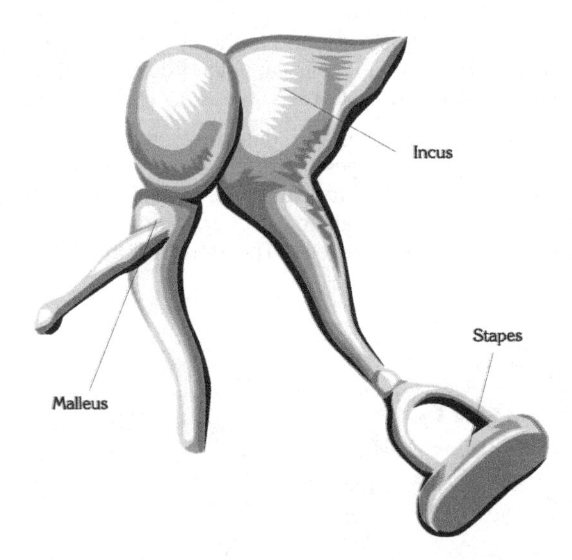

Figure 5.22.: The auditory ossicles are involved in the transfer of sound to the inner ear.

Human ears are made of three main parts: outer ear, middle ear and inner ear. The mechanoreceptors sensitive to vibration are found in the structure of the organ of corti, which is found in the innerear. Auditory nerves connected to receptors carry information detected here to the CNS and the sense of hearing is formed.

C. TRANSPORT SYSTEM (Circulatory System)

The simplest meaning of **"being alive"** is having the ability of energy production. In order to produce energy body cells need food and oxygen. In complex multicellular organisms, [illegible]

When energy is produced as a result of metabolic activities, some unnecessary (waste) products are [illegible] produced and to keep the body in homeostasis these [illegible] The transport system is also responsible for carrying off these substances to the organs which remove wastes. For instance, digested food (nutrients) absorbed from the intestines and carried to all cells of the body by means of the circulatory system. When these nutrients are used, some wastes, including CO_2 and urea, are produced, then they are carried to either lungs or to the kidneys to be removed from the body.

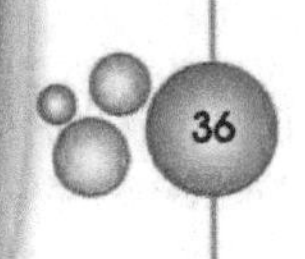

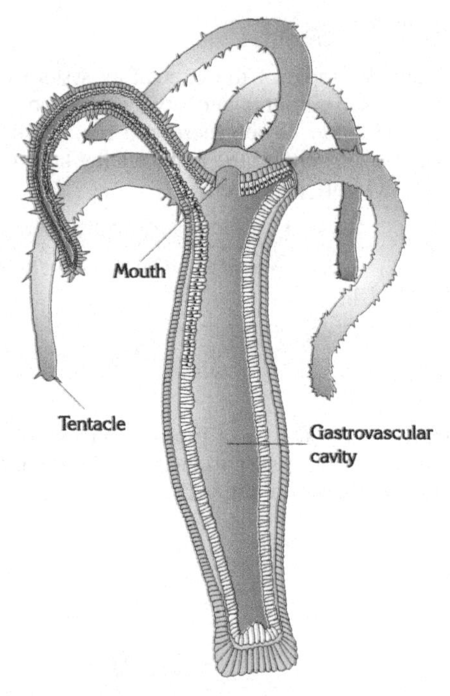

Figure 5.23.: Material exchange in coelenterates occurs through a single orifice.

Among animals there are two main types of circu- [illegible]

- [illegible]
- [illegible]

1. Open circulatory system (OCS)

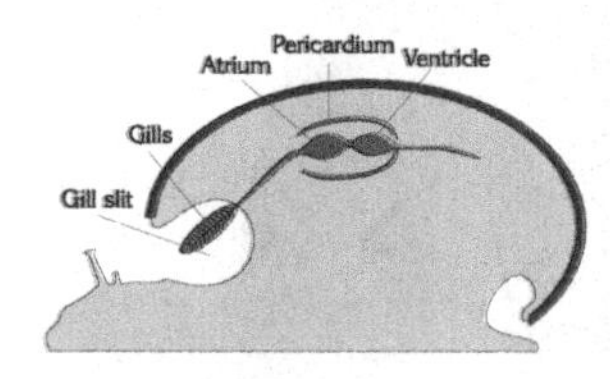

Figure 5.24.: Transport structures in snails

[illegible] Arteries are blood vessels that take blood from the heart to all parts of the body, while veins are blood vessels that bring blood back to the heart from all parts of the body.

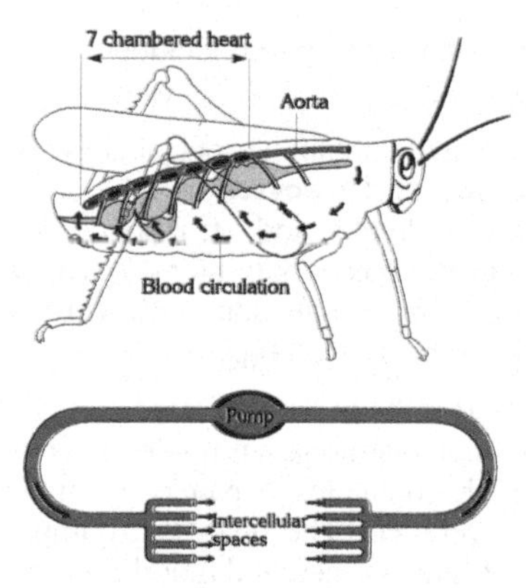

Figure 5.25.: The grasshopper has an open circulatory system. Its blood flows throughout the body cavity as well as in the blood vessels.

In [illegible], blood is pumped into an artery, from the artery to the body cavity, then to veins (after material exchange between blood and body cells) and back to the heart.

2. Closed circulatory system (CCS)

[illegible] Very tiny blood vessels, capillaries are the site of material exchange. Blood never leaves the blood vessels.

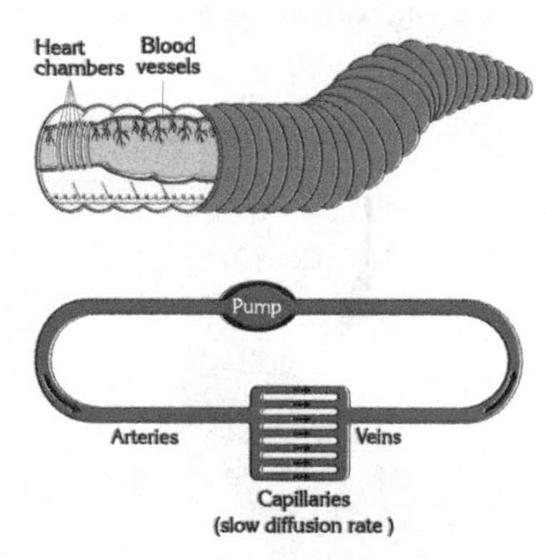

Figure 5.26.:Earthworms are the most primitive organisms with a closed circulatory system.

A true circulatory system includes a heart, blood and blood vessels.

Blood is fluid containing special cells and plasma. Blood cells are **erythrocytes** (red blood cells), that give blood its red color and are responsible for carrying gases, **leukocytes** (white blood cells), which protect the body from infection, and **thrombocytes**, responsible for blood clotting.

The heart is the most important part of the circulatory system. In mammals it is made of two sides, right and left, both containing a pair of chambers, called atrium and ventricle. The heart is a pumping center. From the left side, oxygenated blood is pumped to the cells of the body. Deoxygenated blood, collected from the body via the veins, is sent to the lungs by the right side of the heart.

D. RESPIRATORY SYSTE

Under the topic of the circulatory system it was mentioned that nutrients and oxygen are needed for energy production. When these two react with each other in special parts of the cells, energy and wastes, including CO_2, are produced.

[illegible] (taking in oxygen and releasing CO_2).

[illegible]

Trachea is a system of pipes. Branches of these pipes penetrate all tissues to facilitate the diffusion of gases into all regions of the body.

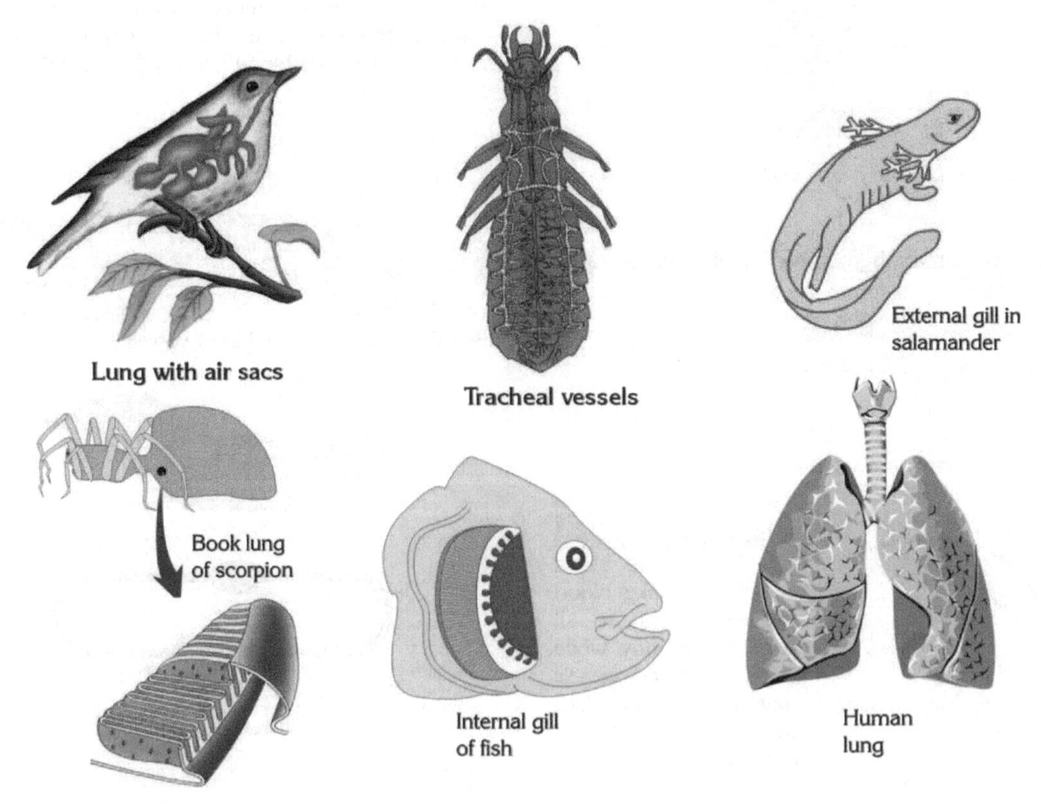

Figure 5.27.: Different types of respiratory structures among animals

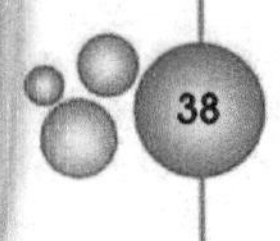

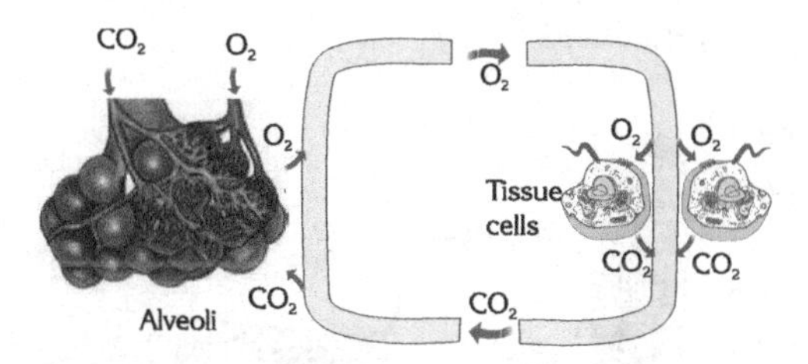

Figure 5.28.: Oxygen and carbon dioxide are exchanged at alveoli and tissues.

Mollusks, echinoderms, crustaceans, fish and amphibians respirate with gills. The most important feature of gills is that they absorb oxygen dissolved in water.

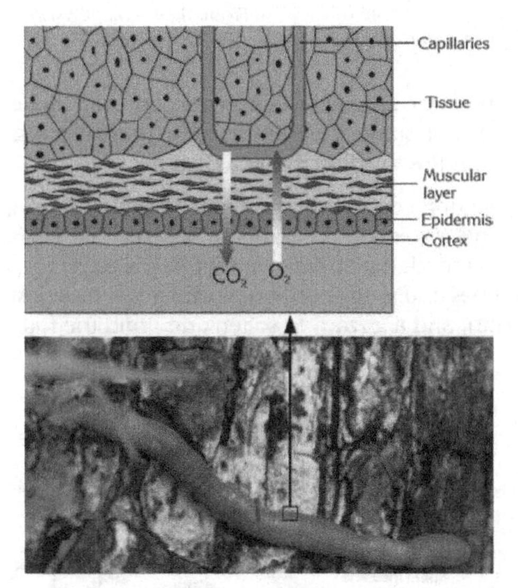

Figure 5.29.: Gas exchange in earthworms occurs by simple diffusion in and out of capillaries under the moist epidermis.

Adult amphibians, reptiles, birds and mammals respire through lungs. The lungs and complementary structures of these organisms have some unique features. Human have two lungs, located in the chest cavity. The respiratory pathway includes the nose, nasal cavity, pharynx, bronchi, bronchioles and alveoli. The sites of gas exchange, alveoli are covered with millions of capillaries. Oxygen absorbed from the alveoli is transported to the body via the circulatory system, and carbon dioxide is brought to the lungs to be exhaled.

E. DIGESTIVE SYSTEM

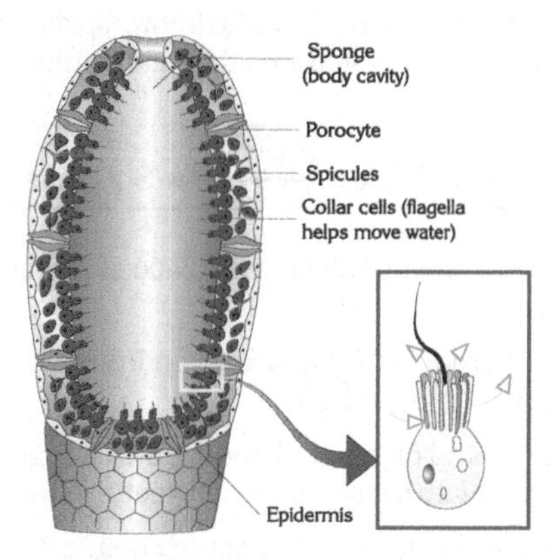

Figure 5.30.: The digestive system of the sponge.

Organisms obtain the energy required for all their metabolic functions, growth, and the repair of their damaged tissues, from food.

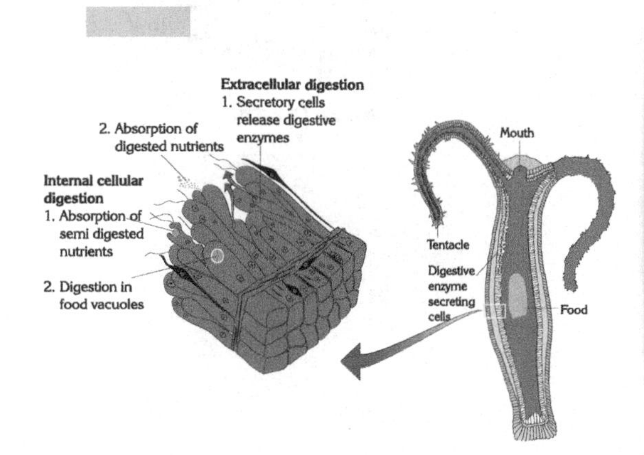

Figure 5.31.: The digestive system of hydra.

In digestion, food is ingested from the medium, then hydrolyzed into its sub-units and absorbed from the digestive canal into the bloodstream. In all animals, vitamins and water pass directly into the bloodstream without any digestion. Carbohydrates, lipids and proteins need to be digested into their components.

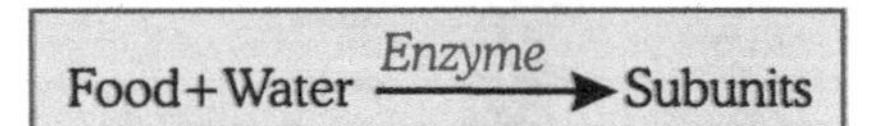

$$\text{Food} + \text{Water} \xrightarrow{\text{Enzyme}} \text{Subunits}$$

The digestion process takes place in certain steps, as follows:

1. Ingestion of food: Food is taken from the external medium into the body.
2. Mechanical digestion: Food is physically ground or chewed into smaller pieces by teeth, in the mouth.
3. Chemical digestion: A series of chemical reactions in which food is hydrolyzed by water and enzymes.
4. Absorption: The final stage of digestion. The subunits of food pass into the bloodstream.

Digestion can take place inside cells. In intracellular digestion, food is taken into food vacuoles and digested. Extracellular digestion occurs when cells secrete enzymes to digest the food outside the body. The digested food then enters the cell through the cell membrane, by diffusion.

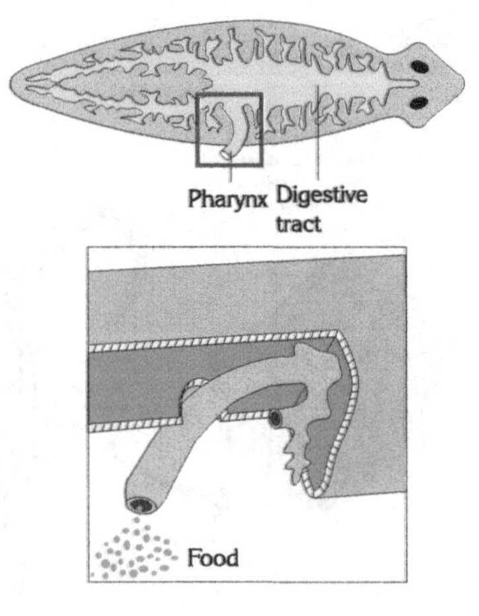

Figure 5.32.: The simple digestive system of a planarian.

Extracellular digestion is seen in invertebrates and in all vertebrates.

In invertebrates

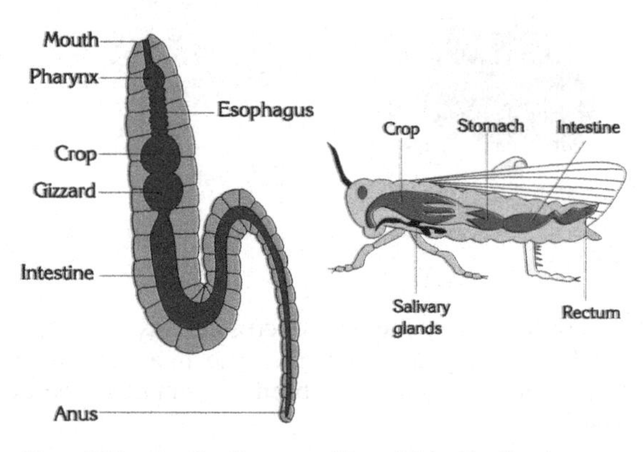

Figure 5.33.: The digestive system of an earthworm.

Figure 5.34.: The digestive system of a grasshopper.

In lower animals there is no particular digestive system, but some specialized cells (primitive organs) carry out this function inside a digestive tract.

Among vertebrates, primitive and well-developed digestive systems and organs are observed. Vertebrates have a mouth, teeth, tongue, esophagus, stomach, and intestines as digestive organs. Birds have a beak instead of teeth, and a gizzard to soften and grind the food.

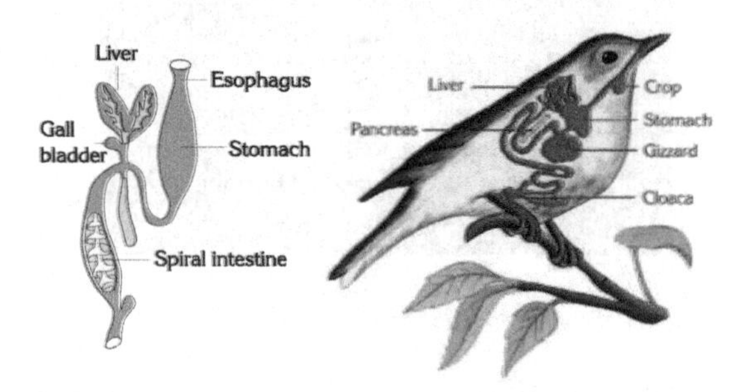

Figure 5.35.: The digestive system of fish.

Figure 5.36.: The digestive system of a bird.

Mammals have the best developed digestive structures among the animals.

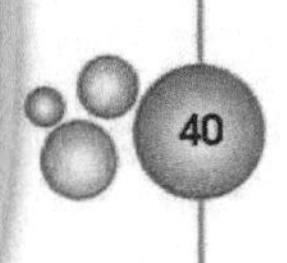

F. EXCRETORY SYSTEM

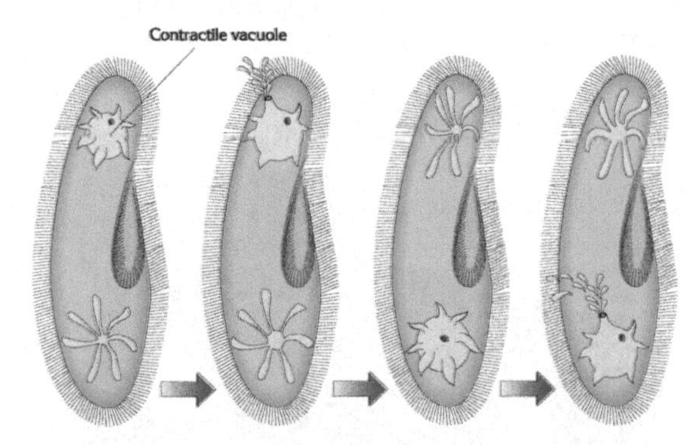

Figure 5.37.: The excretory system of a paramecium, a unicellular organism

Digestion of food and uptake of oxygen were already mentioned in previous sections. by the circulatory system. The in their metabolism In addition to energy, some other,

In this section, the methods of excreting waste products from the animal body will be examined. In brief, excretion is the elimination of metabolic wastes from the body.

EXCRETION IN INVERTEBRATES

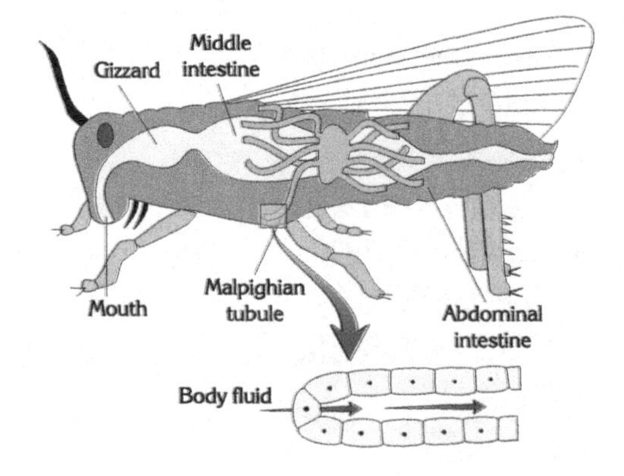

Figure 5.38.: The excretory structures of an arthropod.

Excretion in lower animals, such as planaria, is carried out by protonephridia, which consist of tubules and interconnected flame cells. In earthworms and mollusks, the excretion organs are nepridia that eliminate NH_3 and CO_2 directly from the body via anal openings. Insects excrete CO_2 by means of tracheal vessels, and nitrogenous wastes by way of the Malphigian tubules branched throughout the body.

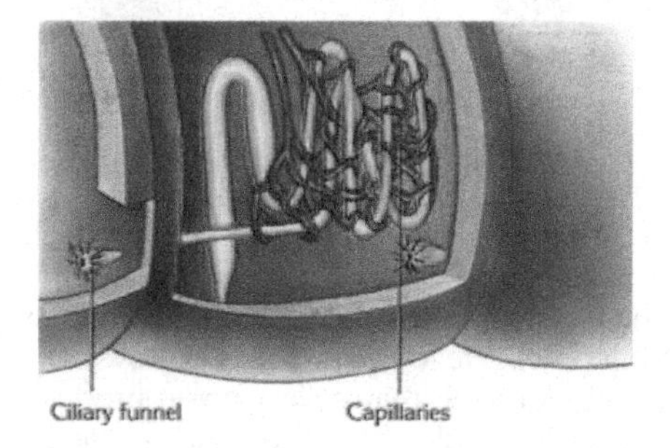

Figure 5.39.: The excretory system of an earthworm.

EXCRETION IN VERTEBRATES

Vertebrates are much more complex than invertebrates, so vertebrates need more developed and complex excretory organs. The

Pronephros

Figure 5.40.: A shark is an example of an animal with a pronephros kidney.

This type of kidney is found in the embryonic stages of all vertebrates, and in adult sharks. It is composed of laterally ordered nephridia, interconnected to each other, with an initial portion that resembles a ciliary funnel.

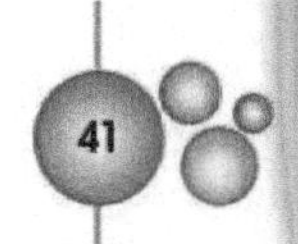

Mesonephros Kidney

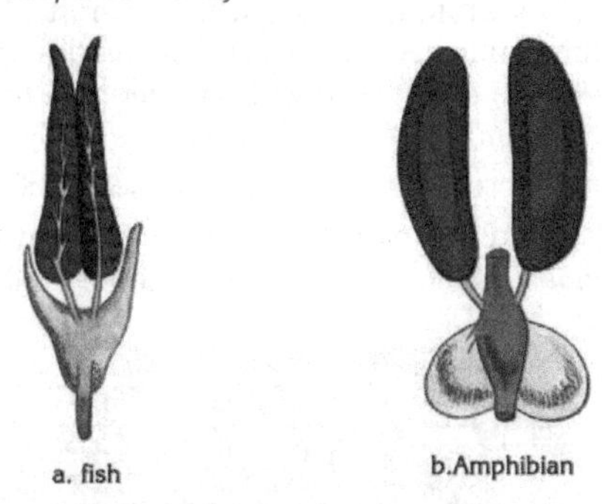

Figure 5.41.:The general structure of the excretory system of a fish and an amphibian.

The mesonephros structure is more complex than the pronephros due to the presence of Bowman's capsules, and glomeruli that collect waste products into the Bowman's capsules. This type of kidney can be seen in the embryonic stages of reptiles, birds, and mammals, and in adult fishes and amphibians.

Metanephros Kidney

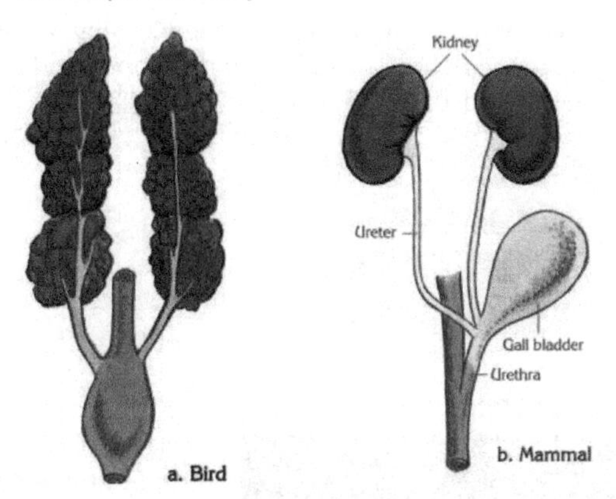

Figure 5.42.: The excretory systems in birds and mammals.

Adult reptiles, birds and mammals, including humans, have metanephros kidneys, found in pairs and located in the dorsal part of the abdominal cavity. Each kidney contains millions of nepridia.

In fish and amphibians, the excretory structure serves in both excretion and reproduction. In birds and reptiles, its function is limited to excretion.

All vertebrates, except mammals, have a single channel for excretion and reproduction. In mammals there is an extra channel for reproduction.

G. LOCOMOTION IN ANIMALS

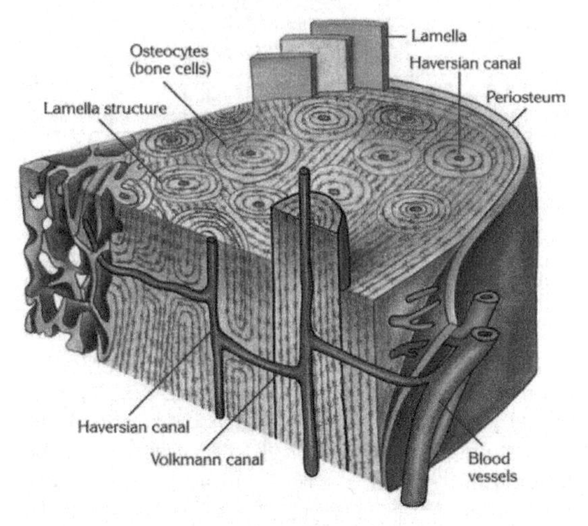

Figure 5.43.: A cut away diagrammatic view of bone, illustrating its structures.

Like all multicellular organisms, animals also need support structures: the skeletal and muscular systems.

[illegible] and support, with the help of the nervous and endocrine systems. [illegible]

- Support the animal body
- [illegible]
- [illegible]
- [illegible]
- [illegible]s
- [illegible]

[illegible]

- [illegible]
- [illegible]

Exoskeleton:

Figure 5.44.: The shells of crustaceans and mollusks function as exoskeletons

An exoskeleton forms the complete outer covering around the animal's body. Structurally, it is composed of organic and inorganic substances. Crustaceans, such as lobsters and many insects, have an exoskeleton [illegible], a complex sugar.

Snails, oysters, and clams have calcium carbonate shells. When these animals grow, they change their exoskeleton by secreting a larger one. The exoskeleton is a hard structure, very useful for protecting animals from enemies, but external skeletons restrict the growth of organisms.

Endoskeleton :

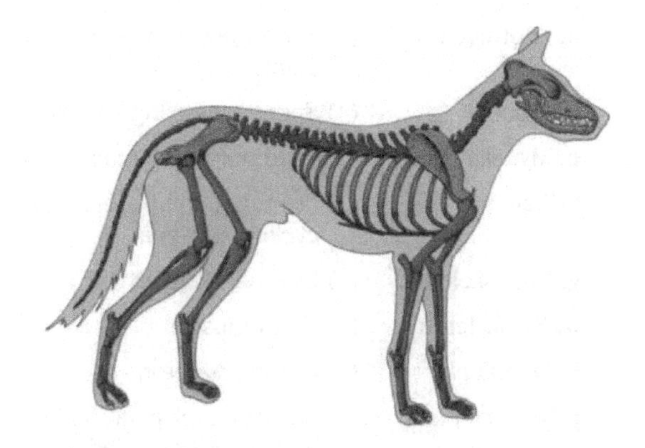

Figure 5.45.: A dog is good example of an organism that has an endoskeleton.

[illegible], starfish, and sponges [illegible] The internal skeleton does not limit the growth of the organism. In most vertebrates, the skeleton is cartilaginous during embryonic development, but later is ossified. However, it persists as cartilage thoughout life in sharks and rays.

[illegible]

1. [illegible] Involuntary muscles found in the internal organs, like the stomach and intestines.
2. [illegible]**:** Voluntary skeletal muscles, attached to bones. They are necessary for locomotion and constitute 40% of total body mass.
3. [illegible]**:** Involuntary muscles that form the heart structure. Stronger and more branched than other muscles.

SELF CHECK

BODY STRUCTURES IN LIVING ORGANISMS

A. Key Terms

Microorganisms	Heterotroph
Root	Stem
Leaf	Virus
Autotroph	Xylem
Phloem	Photoreceptor
Bacteria	Hyphae
Chemoreceptor	Mesonephros kidney
Protists	Mycelia
Metanephros kidney	Endoskeleton
Exoskeleton	Smooth muscle
Cardiac muscle	Striated muscle
Sporangia	Perennial
Biennial	Annual

B. Review Questions

1. Explain the term microorganisms with examples.
2. Compare the structures of bacteria and body cells.
3. List the systems in the animal body.
4. Why are receptors necessary for living organisms?
5. What are the steps of digestion?
6. What are the main differences between OCS and CCS in animals?

C. True or False

1. ☐ Viruses have both DNA and RNA in their nuclear core.
2. ☐ Perennial plants are woody and live more than two years.
3. ☐ Nervous and endocrine systems are the regulatory systems of animals.
4. ☐ Chemoreceptors are sensitive to pressure and temperature.
5. ☐ Smooth muscle and cardiac muscle are involuntary muscles.
6. ☐ Tracheal respiration can be seen among mollusks and echinoderms.
7. ☐ Earthworms are the most primitive organisms to have a closed circulatory system.

D. Matching

a. Cardiac muscle	( ) Plant that completes its life cycle in two years
b. Heterotrophic organisms	( ) A mass which is formed by total hyphae in fungi.
c. Sporangia	( ) Food transporting pipes in plants.
d. Phloem	( ) Spore producing structure
e. Mycelia	( ) Feed on other organisms.
f. Stem	( ) Plant that completes its life cycle in one year.
g. Endoskeleton	( ) External skeleton
h. Exoskeleton	( ) Heart muscle
i. Annual plant	( ) Human skeleton
j. Biennial plant	( ) Body part of plant, either herbaceous or woody.

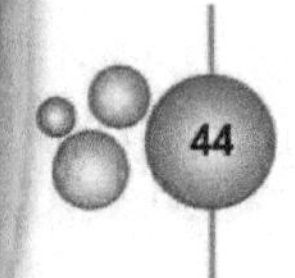

E. Fill in the blank

1. Adult reptiles, birds and mammals, including humans, have .... kidneys.

2. Stomach and intestines are formed by .... muscle, the heart is made from .... muscle, and our bones are covered with .... muscles.

3. In respiration, ... and .... gases are exchanged between an organism and its surroundings.

4. A tree has three main parts, ... , ... and ....

5. Lichens are the symbiosis between .... and ....

6. In higher plants, the .... tubes are responsible for the transport of water and minerals, and the .... tubes are responsible for food transport.

7. The circulatory system in higher animals consists of a .... as pump, ... as agent and ....

F. Multiple choice

1. Most animals stop growing when they become ________.

A) an adult B) old C) strong D) young E) weak

2. ________ do not move by their own will.

A) Bacteria B) Animals C) Plants D) Protists E) Viruses

3. To produce energy, living things__________.

A) breathe B) expire C) respire D) excrete E) grow

4. ______ is throwing the unwanted substances out of the body.

A) Excretion B) Reproduction C) Respiration D) Circulation E) Growth

5. Human beings have ______ senses

A) 4 B) 5 C) 6 D) 8 E) 9

6. __________ make their own food.

A) Living things B) Plants C) Animals D) Fungi E) Viruses

7. Which of the following is an organic substance?

A) Stone B) Water C) Sugar D) Oxygen E) Carbon dioxide

8. Which of the following is not a property of all living things?

A) Reproduction B) Respiration C) Growing D) Photosynthesis E) Excretion

9. Biology is the study of __________.

A) animals B) plants C) worms D) all living things E) bacteria

10. Which of the following is not part of all cells?

A) Cell membrane B) Cytoplasm C) Cell wall D) Nucleus E) Flagella

VI. CLASSIFICATION OF LIVING THINGS "SYSTEMATICS"

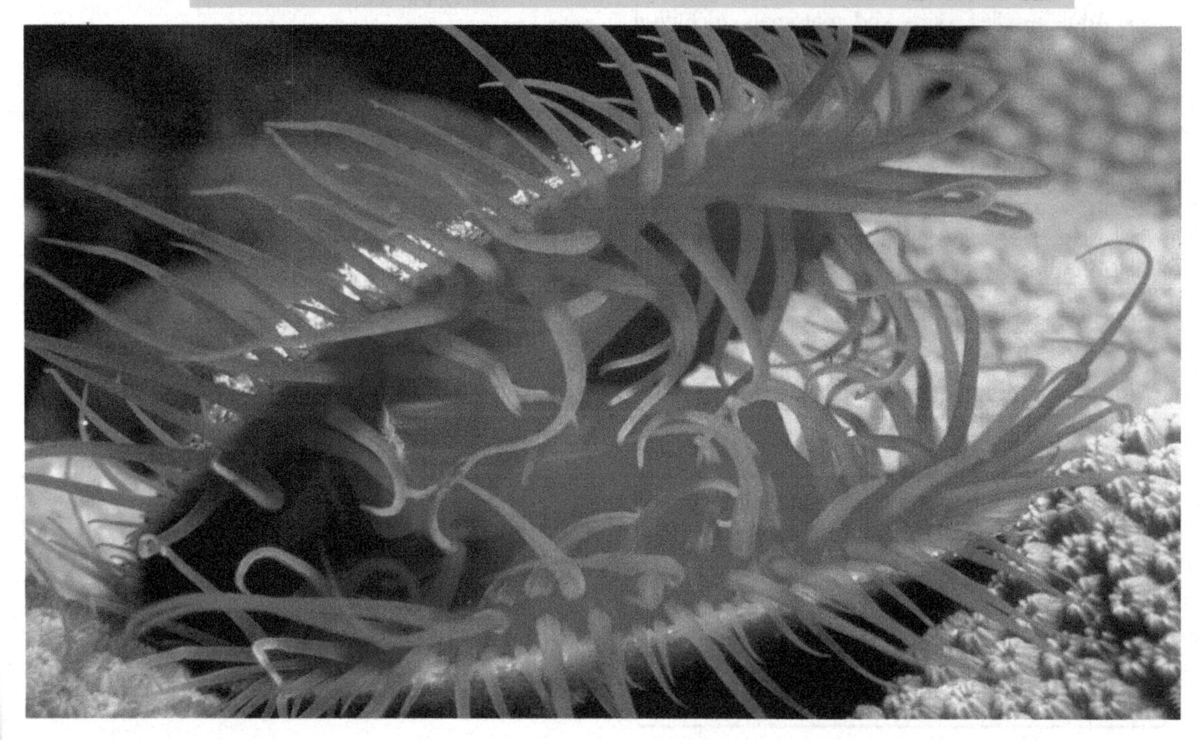

A. [illegible]

Several million types of living things are known to exist today. Furthermore, new species discovered each year increase this number. As millions of organisms may belong to a single species, it is not possible to investigate and recognize the billions of individual organisms.

[illegible] This is called observational classification.

Biologists have used observation methods since ancient times. Although, many classifications are made by observation, some new techniques are used as well.

1. History of Classification

The Greek philospher Aristotle (350 BC) listed only a few hundred plants and animals. Aristotle and his pupils grouped plants as grasses, bushes, or trees, and animals as aquatic or terrestrial. Another philosopher classified animals as useful, harmful or useless. Classification by appearance and similarities in function is called empiric (artificial) classification. This classification is based on observations.

Initially, organisms were classified as plants (motionless organisms) or animals. With the discovery of microorganisms in the 16th century, classification problems became more complex.

Organisms like bacteria, blue-green algae and euglena became problematic for botanists and zoologists.

Use of analogous organs in classification was replaced with use of homologous organs.

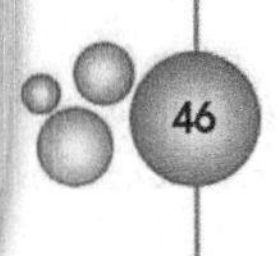

2. Analogous Organs

pro-
vide flight, whereas legs in flies and cats function in walking. But these organs are embryologically different. Therefore we say that butterfies and birds aren't related.

3.

Some similarities may be seen in structures seemingly unrelated in appearence. and embryologically. and similar embryological stages

Before modern classification was developed in the 18th century, several methods had been used. John Ray (1626-1705) tried to unify classification systems and was the first to use the term "species".

He applied his binominal nomenclature method to plants (in 1753) and to animals (1758) in his book *Systeme Naturae*.

4. System of Classification

Phylogenetic systematics used today depends on Linnean systematics and homology. Homology is used in determining the level of relatedness; e.g. bats and humans are in the group mammalia.

The basis of modern systematics is the grouping of organisms according to similarities. In the classification of organisms, the following critera are used: origins, relatedness, developmental stages.

In the binomial system created by Linneaus, species is the basic unit of nomenclature. A species is a group of organisms from the same population sharing the same embryological, morphological, and physio-

Figure 6.1.: Classification units. As you go from species to kingdom the number of organisms and variety increases while from kingdom to species the number of organisms and variety decreases.Organisms with common features form a species

logical features, and are capable of giving birth to fertile offspring when mated under natural conditions.

There are two points in Linneaus' hypothesis:

- There is an ideal type for each species. This ideal type represents the standard features of every single individual of the species.

- The number of species and their types is constant and unchangeable.

According to Linnaean systematics, a species is First is the genus name with the first letter capitalized. Second is the specific epithet (species name) and the first letter is not capitalized. Both are written in italics and in Latin. The reason for this is to have a single name in the scientific world, thus simplifying the study of species. The first name (Canis) shows that these two species are in the same genus.

A group of species similar in some characteristics forms a genus, similar genera form a family, similar families form an order, similar orders form a class and similar classes form a phylum. Similar phyla form a kingdom.

In this system, species level has the largest number of common features and the least number of individuals compared to the higher levels. As you go up, common features decrease while the number of individuals increases. So the fewest common characteristics and the highest number of individuals are found in the kingdoms. The table below shows the classification of several species.

	Ameoba	White Oak	Mushroom	Wolf
Kingdom	Protista	Plantae	Fungi	Animalia
Phylum	Protozoa	Tracheophyta	Mycota	Chordata
Class	Sarcodina	Angiospermae	Basidiomycota	Mammalia
Order	Amecobina	Fagales	Agaricales	Carnivora
Family	Ameobidae	Fagaecae	Agaricaceae	Canidae
Genus	Amoeba	Quercus	Agaricus	Canis
Species	*A. proteus*	*Q. alba*	*A. campestris*	*C. lupus*

5 KINGDOMS

MONERA:

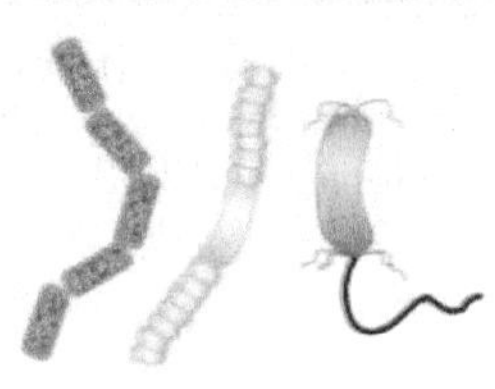

Single celled; Some form colonies; No organelles except ribosomes and Nuclear membrane.Some are motile, some are autotrophs, others obtain food from different organisms.

PROTISTS:

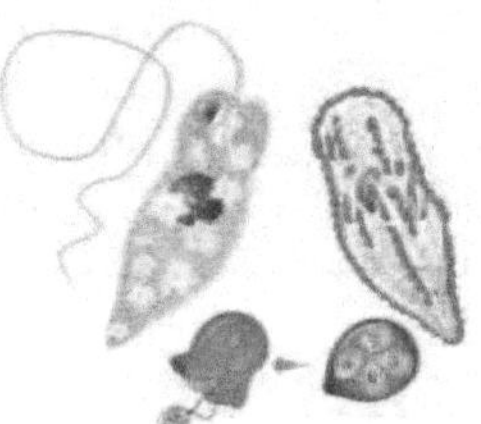

Single or Multicellular; Nucleus and other organelles are present in the cell. Some are motile, some are autotrophs, others are heteretophs.

FUNGI

Multicellular; Nucleus and other organelles are present in the cell. Non-motile and heterotrophic.

PLANTS :

Multicellular; Nucleus and other organelles are present in the cell. Non-motile and autotrophs.

ANIMALS:

Multicellular, nucleus and other organelles are present in the cell. Mobile and feed on other organisms.

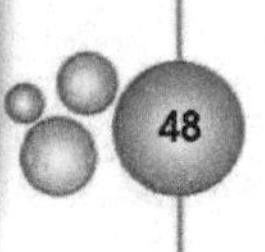

B. Classification of Living Things

[illegible] **(from latin Poison)** [illegible]. Viruses and rickettsias, which do not have the characteristics of living organisms, will also be studied in this unit.

C. [illegible]

Viruses are lifeless outside of living host cells and [illegible] Therefore, viruses are the only organisms in nature to show both nonliving and living characteristics. Whereas normal cells cannot be crystalized, and crystalized cells cannot be revived, viruses can do both.

Viruses are nucleoprotein structures. [illegible]

Features of Viruses:

- They have no cytoplasm or metabolism. They use the metabolism of the cells in which they live.
- Studies with the electron microscope show that viruses may be quite variable in size. [illegible] (10-40 nm), viruses are placed at the bottom of the organization of organisms ([illegible]).
- Information about viruses is obtained from studies of phages and bacteriophages (viruses that feed on bacteria). A bacteriophage virus has DNA inside and a complex protein envelope outside called the **"capsid"**. The viral envelope of the tobacco mosaic virus has been shown to include 158 amino acids and 16 types of proteins.
- Each type of virus infects certain cells. Yellow fever virus infects liver cells.
- Meaning "poison" in Latin, viruses aren't affected by antibiotics but affected only by some physical and chemical factors like pH, radiation, high temperatures and dehydration.

1. Forms of viruses

As mentioned previously, viruses can't live outside the host cell. They are parasites and only function in living cells. They live in plant, bacteria, insects, animal and human cells. Many viruses are pathogens. Furthermore, some cancers are thought to be caused by viruses.

Classification of Animal viruses

Viruses with DNA	Diseases
Adenovirus	Respiratory Infections
Hepatitis Virus	Liver disorders
H.simplex type I	Herpes
H.simplex type II	Genital diseases
Varicella–zoster	Chicken Pox
Epstein–Barr	Certain types of Cancer
Papovavirus	Benign Tumors
Parvovirus	Anemia
Poxvirus	Small Pox

Viruses with RNA	Diseases
Enterovirus	Childhood paralysis, Hepatitis -A, Bleeding Eye disease
Rhinovirus	Cold , Chill
Togavirus	Yellow fever
Paramyxovirus	Measles
Rhabdovirus	Rabies
Coranovirus	Respiratory diseases
Orthomyxovirus	Flu
Arenavirus	Fever
Reovirus	Respiratory and alimentary disease
HTLV I, II	Cancer
HIV	AIDS

Depending on the host organism viruses live in, they are classified as plant viruses, animal viruses and bacteriophages.

a. Plant Viruses

They contain only RNA. Their crystals are needle-shaped. , sugar cane virus, and lettuce virus (Figure 6.2).

Figure 6.2.: Tobacco mosaic virus causes mosaic-shaped areas on tobacco leaves. Diseased (above left) and normal leaves (right) are shown.

b.

These infect animals and humans and cause disease. Their nucleic acid may be RNA or DNA.

c.

Generally they possess DNA. Among these bacteria pathogens, those infecting *Escherichia coli* are studied experimentally.

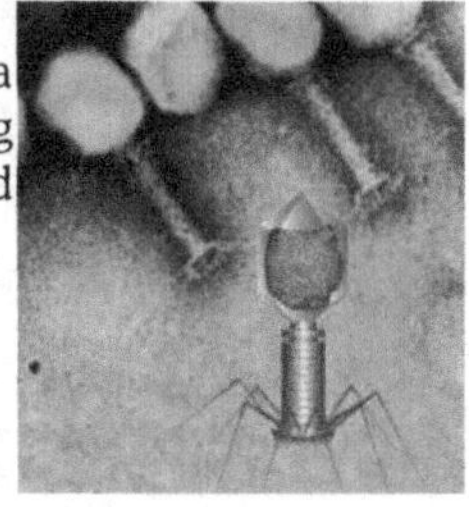

2. Kinds of Viruses :

They are placed in one of three groups generally according to their structure: spherical viruses, rod-like viruses (tobacco mosaic virus) and bacteriophages (Figure 6.3).

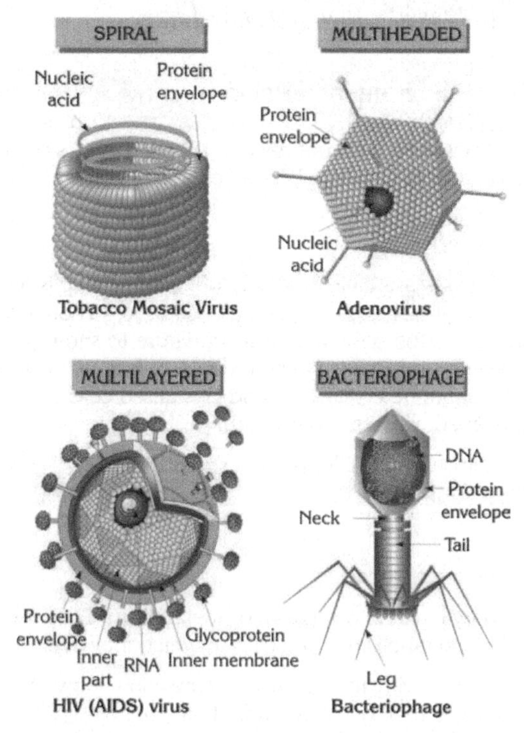

Figure 6.3.: Certain viruses and their structures. As seen in the picture, viruses are made of different combinations of proteins and nucleic acids.

3. (HMW READ)

The protein envelope and nucleic acid give the virus its unique shape. Especially in bacteriophages, extensions of the envelope provide adhesion to the bacterial surface. Special enzymes excreted from these extensions destroy the cell wall. These enzymes are not active during metabolic activities or production of energy, but only in this entrance stage. Finally viral nucleic acid enters the bacterium, while the envelope is left outside.

Viral DNA has two activities in the cell:

- First it takes control of cell metabolism and replicates itself using nucleotides, ATP energy and enzymes. New viruses are formed with DNA and protein envelopes. Thirty minutes after entering, large numbers of viruses have been formed. New

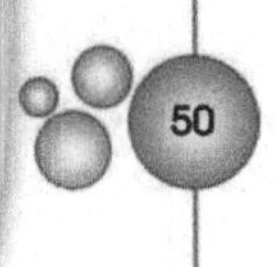

viruses continue to increase until they destroy the cell and exit. This is called "lysis". Free new viruses infect other bacteria.

EXPLAIN!

Phage
Phage attaches to bacterium
Phage injects its DNA into bacterium
DNA of Phage combine with bacterial DNA
Bacterium multiplies by dividing
New Phages develop and get out by bursting bacterium

Figure 6.4.: Viral reproduction occurs only in cells. Viral nucleic acid enters the cell. Viral DNA forms new viruses in the cell. As a result, bacteria burst and viruses are released. Or, viral DNA combines with bacterial DNA causing variation in the bacteria. Viral DNA that has joined bacterial DNA can produce new viruses.

- Second, phage DNA integrates with bacterial DNA. This integration is totally harmless and it doesn't affect bacterial metabolism. Replication of phage DNA continues with that of the bacterium, so new viruses are produced continuously (figure 6.4).

D. Rickettsias

Rickettsias are more advanced than viruses. In a sense, they are intermediate between viruses and bacteria. They are smaller than bacteria, and spherical or rod shaped. They live parasitically inside cells. They can't be filtrated and they are barely visible under light microscopes.They can't survive outside the cell. Their nucleic acids are DNA or RNA. They possess enzymes of bacteria and reproduce with amitosis.

DISCOVERY OF VIRUSES

In 1892, the Russian scientist Iujanolusky found that viruses could live in organisms. Viruses were disvored with the invention of the electron microscope in 1950. Robert Koch, Louis Pasteur and other bacteriologists found bacteria to be pathogens for many diseases. But some diseases were surprising to them because they could not find any bacteria in the infected organisms. The first such disease was encountered in tobacco leaves. The leaves of the infected plant were covered with mosaic-shaped spots and, therefore, the disease was called "Tobacco mosaic disease". An extract from leaves was filtered through Chinese filter, and all bacteria were captured. The extract was then applied to the healthy leaves and seen to cause the disease. Since there were no bacteria in it, the unknown factor in this extract was named Beiyernick. This factor later was identified as tobacco mosaic virus. American microbiologist W. M. Stanley succeeded in isolating the virus. At the beginning of the 20th century many other pathogenic viruses were identified.

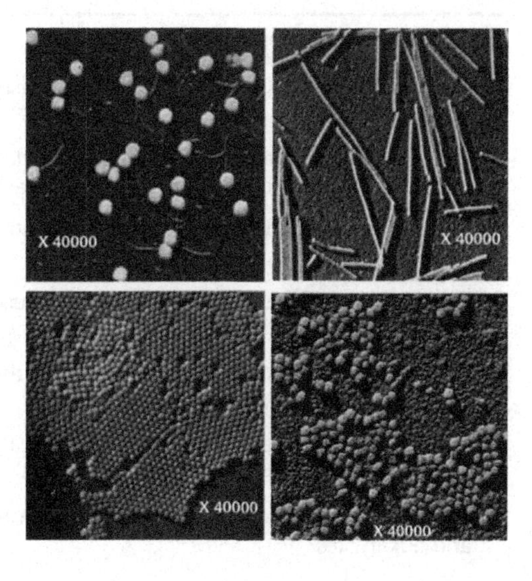

Rickettsias live in the gut of blood-sucking animals [illegible] like spotted fever and typhus. Rickettsias are able to infect domestic animals (sheep, goat, cow, pig, and dog) and cause death.

Protection is provided by a number of special methods. Eliminating rats and other rodent vectors should be an alternative method.

SELF CHECK

SYSTEMATICS

A. Key Terms

Classification	Lysis
Analog organ	Monera
Homolog organ	Protists
Caspid	Fungi
Virus	Plants
Binomial naming	Animal

B. Review Questions

1. List characteristics used to separate organisms into kingdoms.
2. Describe the characteristics of each kingdom and give an example of an organism from each.
3. List factors that result in changes in classification systems.
4. Describe the differences that separate viruses from living things.
5. Describe two characteristics of viruses that make them similar to living things.
6. How could many cells contain copies of the nucleic acid from one virus?
7. What are the differences between prokaryotes and eukaryotes?
8. Why is binomial nomenclature not sufficient for classifying organisms today?
9. What is the difference between analogous and homologous structures?
10. The following organisms all belong to the same family. Which two are most closely related: *Gazella thompsonii, Antilope cervicapra, Gazella arabica*? Explain the reason for your answer.
11. Explain how species maintain their distinctiveness.
12. What condition is necessary for organisms to be considered members of the same species?
13. Which Latin name most specifically describes human beings in taxonomy?
14. Write the reasons why viruses are not accepted as true living things.

C. True or False

1. ☐ Analogous organs are functionally same, but embryologically different.

2. ☐ Meaning of virus is "poison" in Latin.

3. ☐ Many viruses are not pathogens

4. ☐ Viruses are discovered in 1950

D. Matching

a. Binominal nomenclature	( ) Seedless plants
b. Taxonomy	( ) Animal that can be divided by any number of planes passing through the body axis.
c. Vertebrate	( ) Animals with backbones
d. Radial symmetry	( ) Flowering plants
e. Angiosperms	( ) Science of classification.
f. Fern	( ) Double naming system.

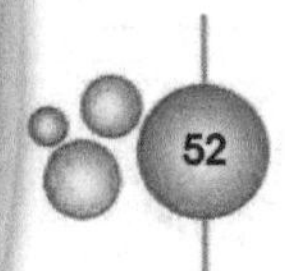

E. Fill in the blank

1. .... has the smallest number of individuals but has the highest similarity and relatedness among its members.

2. ....kingdom has properties that are similar to both the plants and animals.

3. According to their genetic material, there are two types of viruses. These are .... viruses and .... viruses.

4. The meaning of virus in Latin is ....

5. As you go up from .... to ...., both the number of organisms and variation increase.

6. In modern classification used today, each organism has ....names irrespective of its kingdom.

7.. Of what use to a taxonomist is knowledge of the amino acid sequences of various organisms?

8. The science of describing, naming, and classifying organisms is ....

9. The kingdom that includes the algae is ....

10. Related classes are grouped together in the same ....

11. Kingdom .... consists of decomposers such as molds and mushrooms.

12. Closely related genera may be grouped together in the same ....

13. Give examples of homologous structures.

14. Give examples of analogous structures.

15. In which kingdom would you classify each of the following?

(a) oak tree, ...

(b) amoeba, ...

(c) *E. coli* (a bacterium), ...

(d) tapeworm, ...

F. Multiple choice

1. Which one includes all the others ?

A) Species B) Class C) Pyhlum D) Order E) Genus

2. An independent organisms is discovered that does not contain a nucleus. In all likelihood, it would be classified in the kingdom of ....

A) monera B) protista C) fungi D) plantae E) animal

3. The scientist who gave us the binomial system of classification was ....

A) Linnaeus B) Mendel C) Mendeleeff D) Ehrlich E) Metehnikoff

4. Which of the following is not chracteristic of species?

A) Named by two name

B) Unit of classification

C) They have same chromosome number

D) They have same genes

E) The smallest unit of classification.

5. Which is the main reason for classifying living things ?

A) For observing more life types

B) For finding the relationship among living things

C) For finding more life types

D) For observing to development of living things

E) For providing observation and learning facilities

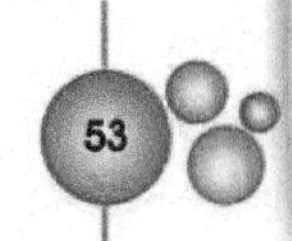

VII. KINGDOM MONERA

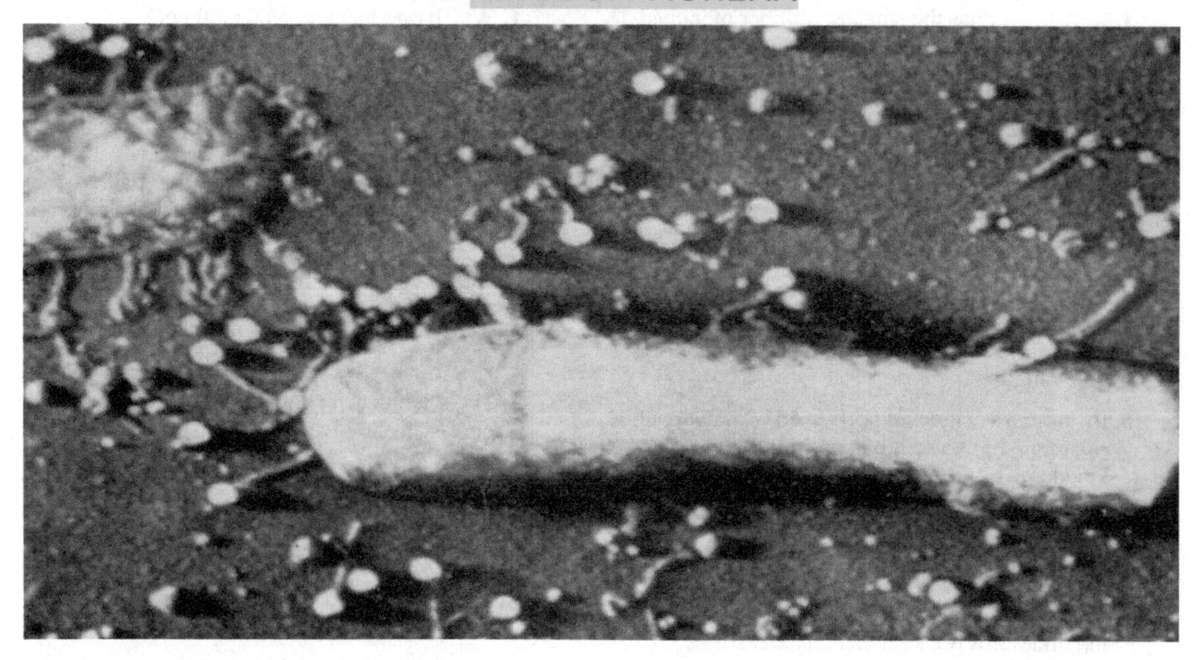

, but some are multicellular in appearence. cell organelles such as , mitochondria , plastids etc. except for stuctures like flagellum seen in some.

organisms. asexual, division or . There are several types of genetic material transfers like conjugation, transformation, tranduction, and plasmid transfer. Motility is provided by cytoplasmic flow, flagella or gliding. Examples are bacteria and blue-green algae.

A.

. A bacterium is about 1/1000th of a eucaryotic cell in volume.

Bacteria are the most numerous organisms in the world and found almost everywhere. They can live 5m below ground, in all waters, and in the body of any organism. Bacteria are the main subjects in biological studies. Especially from the molecular studies of *E.coli*, more detailed information about the organismal and cellular nature of life has been obtained. Today genetically engineered bacteria are used for treatment of human diseases, eg. to produce missing proteins like insulin or interferon.

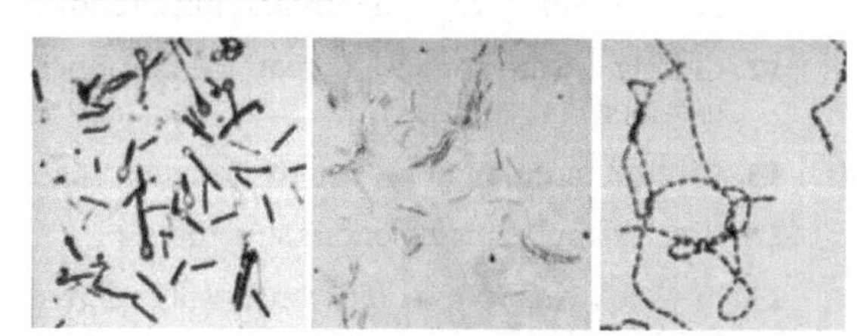

Figure 7.1.: *Some views of bacteria under light microscope.*

1. Structure of bacteria **hmw: read**

Bacteria have no nuclear membrane, chlo

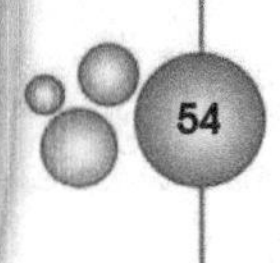

Cytoplasmic membranes of some bacteria carry respiration enzymes, photosynthetic enzymes and receptor proteins. Enzymes responsible for oxidative phosphorylation in the mitochondria of eucaryotic cells are found in mesosomes formed by membrane extensions in aerobic bacteria.

Outside the membrane there is a cell wall different from the cell wall of plants. The structural unit in the moneran cell wall includes diaminopimelic acid (an amino acid) and muramic acid (a glucose derivative). Additionally some protein and oil molecules are found in the cell wall. , made up of polysaccharides, from phagocytosis and absorption to cell surfaces. (Figure 7.2).

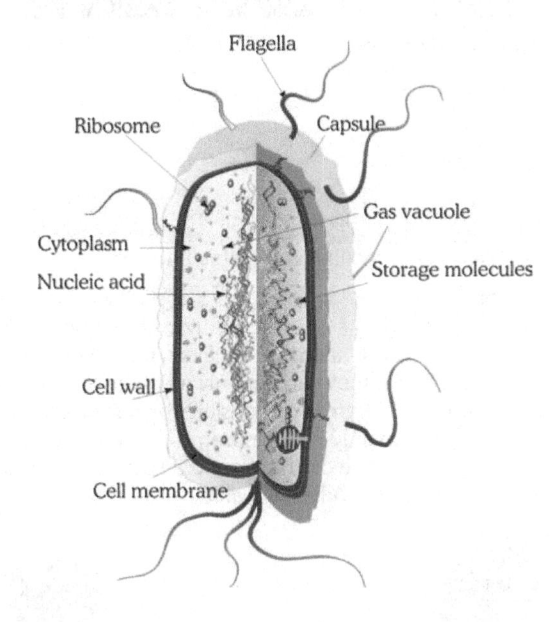

Figure 7.2.: Structure of bacteria. Bacterial cells are prokaryotic. Their nucleic acids and other structures are found in the cytoplasm.

Many bacteria have extensions called flagella. These are involved in movement and signal recognition. They may be positioned as bipolar, unipolar or throughout the cell wall. They are generally found in spiral or rod-shaped bacteria. Spherical bacteria do not have any flagella or active movement.

The cytoplasm of bacteria and that of eucaryotic cells are alike. Inside, glycogen, oil and protein particles are found. DNA and RNA molecules are found in the cytoplasm as there is no nuclear membrane. The area in which nucleic acids are concentrated is called thc nuclcoid. There are also gas vacuoles in the cytoplasm.

Bacteria are able to travel long distances via dust particles or water vapor.

2.

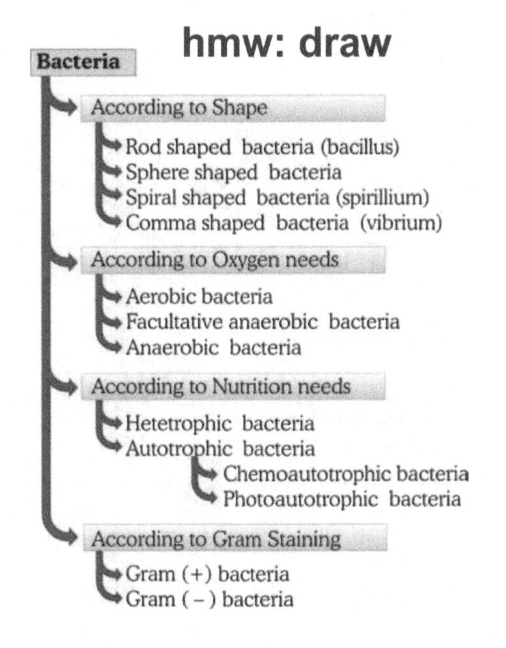

a. S

- : These are rod or needle-shaped. Yogurt bacteria are in this group and possess some enzymes that aid digestion.
- : These are spherical in shape and live solitarily or in

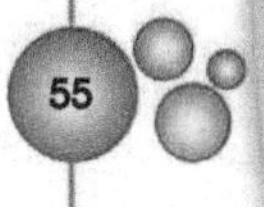

colonies which are identified according to number and shape. Bacteria like a single sphere are called monococcus; those like a double chain are called diplococcus; those like a chain in shape are called streptococcus; those like a cluster of grapes are called staphylococcus. Eg, *Neisseria gonorrhoeae* is a diplococcus bacteria and causes the disease called gonorrhea.

- **Spiral or curved bacteria:** These are spiral shaped and have flagella on both sides. Most are saprophytic. *Treponema pallidum*, which causes syphilis, is a representative of this group.
- [illegible]: These bacteria have one end curved (like a comma mark) while the other end has flagella (Vibrio comma) (Figure 7.3).

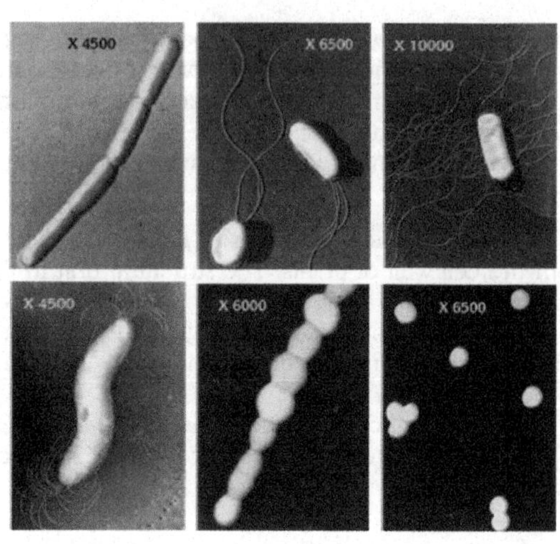

Figure 7.3.: Figures and electron microscope photos of bacterial types. As seen above, bacteria can be rods, spheres, spirals, flagellated, unflagellated or other shapes.

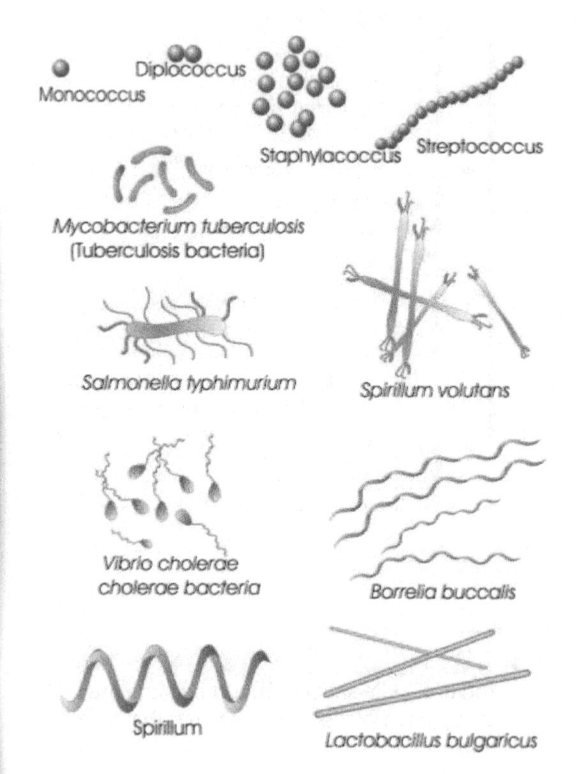

b. [illegible]

[illegible], or facultative aerobic bacteria.

- [illegible] Their respiration is aerobic (as in plants and animals). Because they lack mitochondria, they can get only a limited amount of energy. Enzymes responsible for respiration are found mainly in mesosomes. These bacteria generally live on the ground and on or near the surfaces of bodies of water.

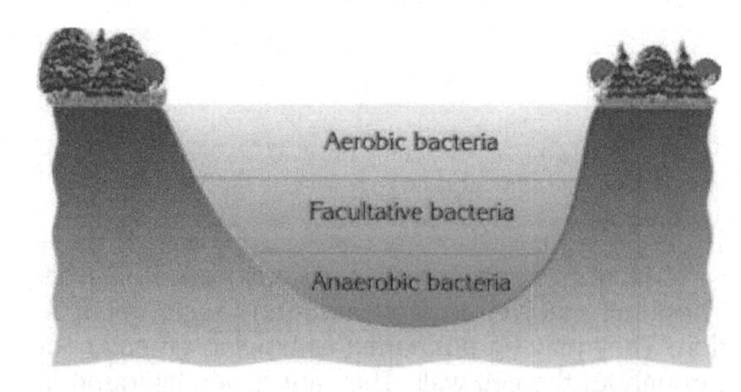

Figure 7.4.: Bacteria, according to respiration type, live in different types of soil and water. Aerobic bacteria live on or near the surface, facultative bacteria live in the middle layers and anaerobes live in deep layers.

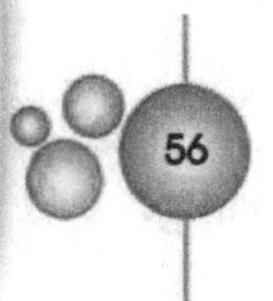

- **Anaerobic bacteria:** [illegible] Waste products are generally alcohol, lactic acid or acetic acid. Their carbohydrate metabolism is called fermentation, while protein and amino acid break down is called putrification.

  Anaerobic bacteria can't survive in oxygenated conditions. Therefore, these organisms are found in deeper layers of soil and at the bottoms of lakes or seas.

- [illegible] and can be found in all layers of soil or water.

c. Nutrition Types

Bacteria are divided into autotrophs or heterotrophs on the basis of feeding behavior.

- [illegible]

  - [illegible] They lack digestive enzymes. For this reason [illegible] in order to survive. As the human body has such matter (glycerol, glucose, fatty acids and amino acids) in the digestive system, blood and cells, [illegible]

    Pathogenicity depends on dispersal ability, reproductive rates, and efficiency of toxins produced. Toxins include exotoxins and endotoxins. Exotoxins are proteins and highly efficient. Endotoxins are produced inside bacteria and are exposed to cells after the destruction of the bacteria. Their pathogenicity is far less than that of exotoxins. Pathogens destroy the cells in which they live, or secrete one of several toxins that cause disease. Some diseases may result in the death of the host organism. When the host dies, the bacteria lose their food and home. The bacteria that are best adapted to the host's metabolism do not cause death. So the most advanced type of parasitism is that which does the least harm to the host organisms.

  - [illegible] to simpler compounds. They convert organic matter to inorganic. During this process, they obtain nutrients and enrich the organic content of the soil. Soil with an abundance of organic content is called humus. Saprophytic bacteria have an important role in recycling carbon and nitrogen, both of which are essential components of living organisms. Without saprophytic bacteria there would be no cycling of nutrients in the environment. As a result of decomposition of organic matter, foul-smelling nitrogen and sulphur compounds are released.

    Many branches of industry make use of the chemicals released during bacterial metabolism. Ethyl alcohol, acetic acid, acetone and butyl alcohol are examples of chemicals produced by special bacteria. Other areas where saprophytic bacteria are used include the leather and tobacco industries, production of cheese, pickles, yogurt, and treatment of industrial wastes.

- [illegible] and convert inorganic chemicals to organic matter. [illegible] identified according to energy source: [illegible]

  - [illegible] In their cytoplasm they have pigments like chlorophyll but they don't have chloroplasts.

  - [illegible]

Methods of Bacterial Infection

Bacteria invade the human body in food and water via the mouth, in the air via the nose and mouth, via the skin, or via the sex organs. In plants, scars, stomata and flowers are infection areas. Both in human beings and plants insects are the main vectors.

d. Gram staining

Because of structural differences in the cell wall, According to this phenomenon, bacteria are divided into two groups:

- They have thick cell wall layers. The outer layer is thicker and formed from peptidoglycan, while the inner layer is composed of lipids and carbohydrates.
- Their cell wall consists of a thin peptidoglycan layer sandwiched between a double layer of plasma membranes.

3. Reproduction In Bacteria **hmw: read**

Bacteria reproduce asexually. Under ideal conditions, they divide every 20 minutes and increase geometrically. Theoretically the clone of a single bacterium would produce 2000 tons of bacteria in just 254 hours. But this never happens because bacteria quickly consume the oil, water and nutrients in their environment. Additionally accumulating waste products like alcohol, acids and other chemicals limit growth and can eventually kill the entire colony. So the growth of bacterial colonies is controlled automatically. Cell division in bacteria isn't true mitosis but is suggestive of the mitosis seen in higher organisms (Figure 7.5).

DISCOVERY OF BACTERIA

Scientists started to discover the world of microorganisms with the invention of microscopes in the late 1600s and early 1700s. Anthony Van Leeuwenhoek made high-quality lenses and constructed the first microscope. In 1627 he published his observations of small organisms seen in water droplets. His knowledge and classification is even applied today (eg. classification of bacteria according to shape). In the 19th century German botanist F.J.Cohn first studied the structure of the bacterial cell wall.

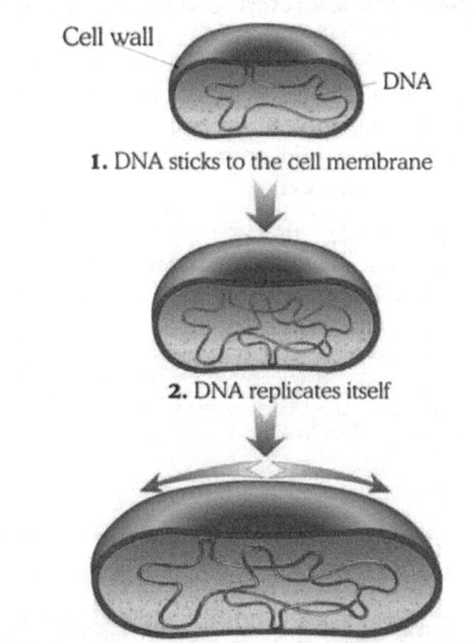

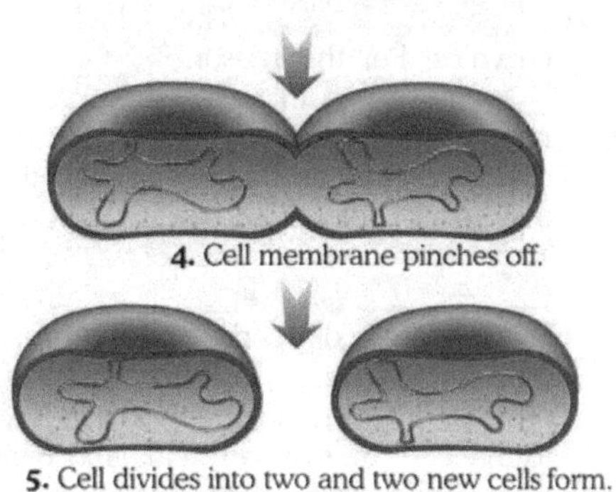

3. Cell membrane stretches and pulls the DNA to the poles

4. Cell membrane pinches off.
5. Cell divides into two and two new cells form.

Figure 7.5.: In bacteria, the DNA is single-stranded and circular. During cell division the DNA replicates itself while attached to the cell membrane. The cytoplasm furrows and division is completed.

Conjugation, transformation and transduction aren't types of sexual reproduction but they do increase genetic diversity. In conjugation, genes are transferred between cells through a temporary cytoplasmic bridge. In transformation, bacteria take up fragments of foreign DNA from their environment and integrate these into their own genome. Transduction is the viral transfer of genetic material.

HMW:READ

4. *Endospore Formation in Bacteria*

Some bacteria form endospores as a protection These are oval or spherical structures found near the cell membrane or in the cytoplasm. During spore formation, the cytoplasmic content decreases to form the spore, while other parts are discarded. A thick protective layer forms around the spore. Since this spore is formed inside the cell it is called an endospore. The spore is quite resistant to extreme physical conditions like drought, high temperatures, radiation or freezing.

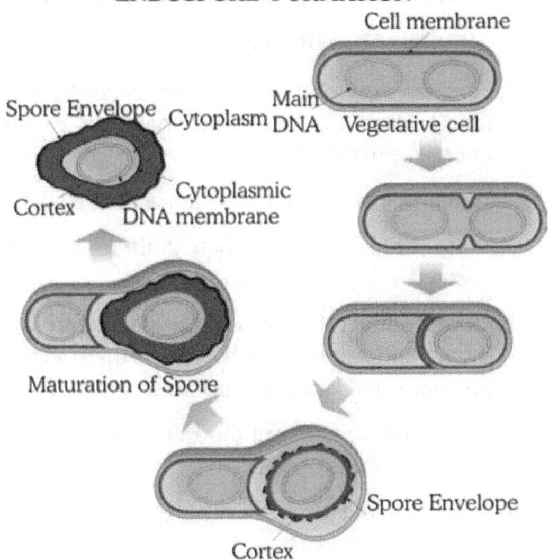

Figure 7.6.: To survive unfavorable conditions, bacteria form a structure (endospore) with a tough coat covering the cytoplasm.

number of spores and bacteria would be the same. It is, however, a dormant state and is seen commonly in bacillae, to a lesser extent in spherical bacteria, and never in spiral bacteria.

B.

Although called , blue-green algae are bacteria and Their cell walls contain cellulose and pectin, and are covered with mucous secreted by the cell. The cytoplasm is colorless in the center where the DNA is found. Peripherally, a colorful layer consisting of and RNA (Figure 7.7).

Photosynthetic blue-green algae are more like photosynthetic bacteria than green algae. However, unlike photosynthetic bacteria, blue-green algae produce oxygen. They are the most primitive algae and are microscopic in size. They fix atmospheric nitrogen and convert it to other forms. They survive bad conditions by endospore formation.

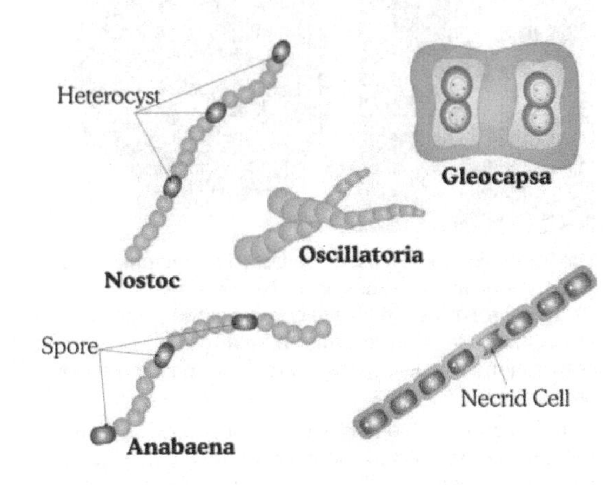

Figure 7.7.: Schematic views of blue-green algae. These photosynthetic bacteria live on rocks or in mud, moist soil and in water.

Blue-green algae are common throughout the world especially on rocks, in moist soil and in water. An abundance of them changes the color of water and produces a bad taste and smell. Internal gases keep aquatic species near the water surface. Some genera are found in spas. They reproduce asexually with spores. Sexual reproduction is unknown in this group. Examples of this group include nostoc, gleocapsa and ascillatoria.

PREPARATION OF BACTERIA CULTURE

Bacteria can easily reproduce in lab conditions. For this, a medium that will nourish the bacteria is prepared. The medium can be broth or agar. If you open a sterile test tube of broth and wait a few days, you will observe turbulence in the tube. Serial dilution can give an idea about the number of bacteria. In the same way, if you open a petri dish, inject some bacteria culture and then close the dish, you will observe some colored spots. These spots show the existence of bacteria.

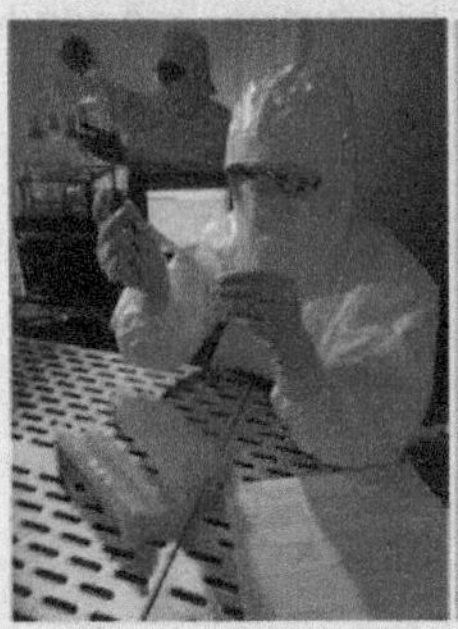

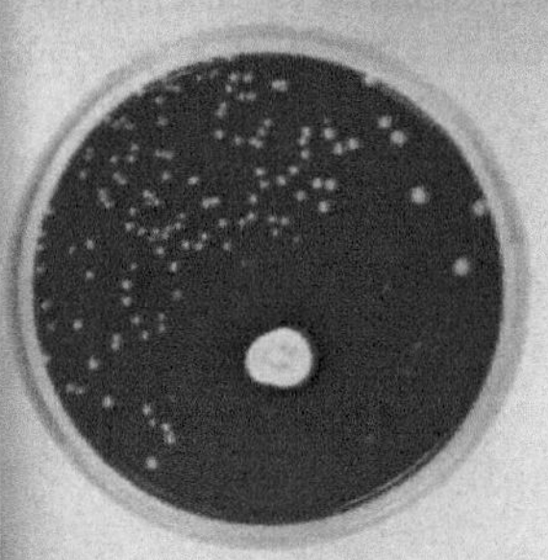

The most commonly used medium for bacteria is agar. Agar is a gelatin-like substance that dissolves in water at 90 °C and solidifies at 40 °C. According to the bacteria to be grown, necessary nutrients are added to the agar during preparation. The agar is inoculated with bacteria to form pure colonies.

To this culture antibiotic is added to see how bacteria are affected. When affected, no colony forms. If they survive, they are demonstrated to be resistant to the antibiotic. Bacteria can be observed under a microscope. For this a bacterial culture, physiological water, an inoculating loop, a slide, a cover slip, and a Bunsen burner are needed.

First, sterilize the slide by holding it over the flame. Put a drop of water on the slide, and then your specimen. Dry the slide in air and stain it. Use an oil immersion objective to observe your bacterial culture.

SELF CHECK

MONERA

A. Key Terms

Prokaryote	Saprophyte
Aerobic	Parasite
Anaerobic	Asexual
Photosynthetic	Transduction
Heterotrophic	Conjugation
Chemosynthetic	Transformation
Facultative	Endospore

B. Review Questions

1. How do aerobic bacteria release energy from food? How do anaerobic bacteria release energy from food?
2. What are two ways that bacteria help plants?
3. Write what would happen if there were no bacteria.
4. Before refrigeration, people preserved food by drying and salting it. Why would these actions prevent some kinds of food from spoiling?
5. Learn and write about the *E. coli* bacteria. Are they dangerous or not?
6. What are the distinguishing characteristics of bacteria?
7. Why are bacteria mostly used in genetic engineering and scientific research?
8. Draw a typical bacterial growth curve and label the different phases. Discuss the factors involved in each phase.
9. In what ways are bacteria useful in the dairy industry?
10. Which diseases are caused by bacteria?

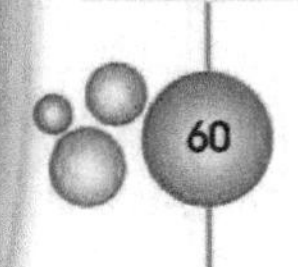

11. What types of bacteria have been used as bio-terror weapons?
12. What are some of the extreme environments in which bacteria are found?
13. What is *E. coli*? Is it always useful?
14. What causes botulism? What is this disease?
15. How is botulinum toxin used in a positive way?
16. What are plasmids?
17. Compare bacteria to other unicellular organisms. How do they differ? How do they compare to viruses?
18. Explain how bacteria change and adapt to new environments.
19. What is a plasmid and what is its importance for genetic engineering?
20. What is a pathogen? Are all bacteria pathogens?
21. Give examples of human diseases caused by bacteria.

C. True or False

1. ☐ Kingdom Monera includes bacteria and blue-green algae.

2. ☐ Bacteria has eukaryotic cell structure.

3. ☐ Bacteria reproduce sexually.

4. ☐ Gram positive bacteria do not absorb gram stain.

D. Matching

a. Flagella	( ) oxgen dependent
b. Conjugation	( ) causing disease
c. Anaerobic	( ) oxygen independent
d. Aerobic	( ) cytoplasmic extensions
e. Autotroph	( ) producing own food
f. Fern	( ) gene transfer through cytoplasmic bridges.

E. Fill in the blank

1. All bacteria are classified in the kingdom ....
2. Bacteria may survive unfavorable conditions by forming....................
3. .... bacteria need light for life.
4. The ....of the bacteria is sensitive to gram staining.
5. .... bacteria feed on the remains of other organisms.
6. Bacteria are a source of .... such as streptomycin and penicillin.
7. Bacteria reproduce by..........., one cell being able to give rise to over 4 x 10^{21} cells in 24 hours.
8. Spherical bacteria are referred to as .... , rod-shaped bacteria as .... , and helical bacteria as ....
9. In ...., genetic material is transferred from one bacterium to another.
10. Some bacteria produce and secrete disease-causing poisons known as ....
11. The methanogens are .... that produce methane from carbon dioxide and hydrogen.

F. Multiple choice

1. Bacteria, blue green algea and viruses belong to kingdom

A) Monera B) Protista C) Fungi D) Plant E) Animals

2. Which of the following organisms are important in recycling organic materials ?

A) Virus B) Bacteria C) Blue green algea D) Rhizopoda E) Sporozoa

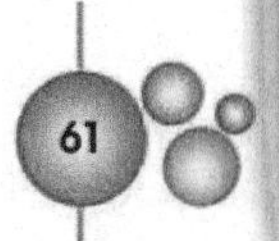

VIII. KINGDOM PROTISTA

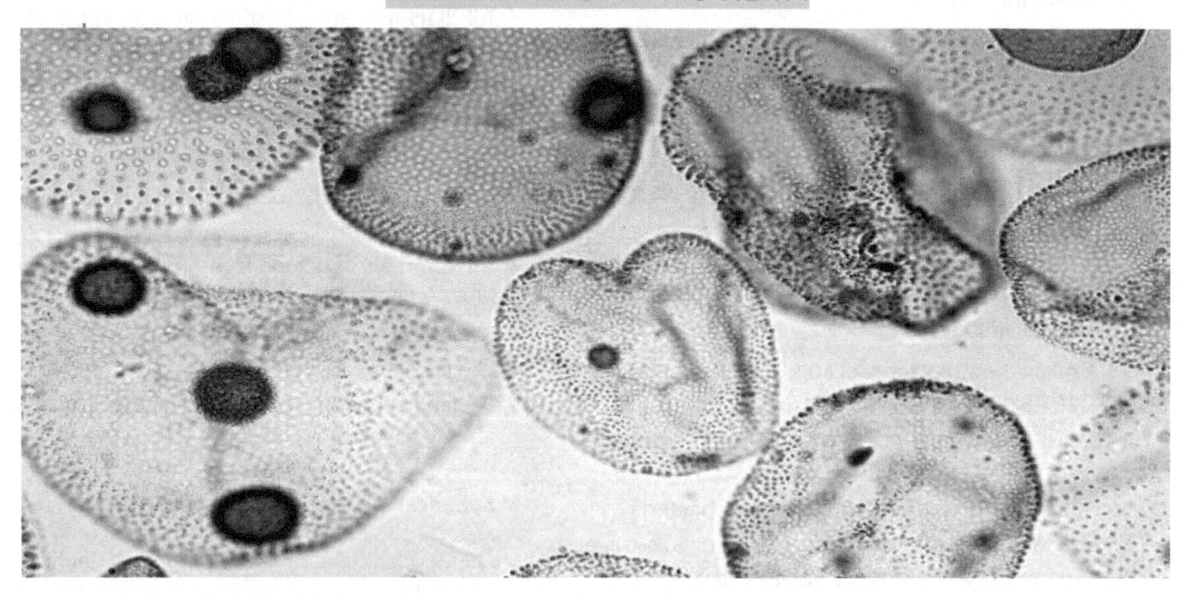

Protists have various sizes, shapes, methods of feeding and reproducing, and life histories. Although their features are obviously different from those of monerans, there are several ways that some protists resemble plants or fungi. Size is generally microscopic, but examples They may be unicellular, colonial, or even multicellular, in which a primitive structure is observed while tissue level organization is absent.

floating freely or attached to a surface, eg. rocks Terrestrial protists are mostly found in moist soil or debris.

in some members. Mutualistic or parasitic forms are common and parasitic protists can cause diseases in plants and animals.

They are motile in certain periods of their lives and (amoeboid movement),

Despite the absence of specialized genitals, asexual and sexual reproduction are observed in the group.

- Phylum Rhizopoda
- Phylum Foraminifera
- Phylum Actinopoda
- Phylum Zoomastigina
- Phylum Ciliophora
- Phylum Apicomplexa

- Phylum Dinoflagellata
- Phylum Bacillariophyta (Diatoms)
- Phylum Euglenophyta
- Phylum Chlorophyta (Green Algae)
- Phylum Rhodophyta (Red Algae)
- Phylum Phaeophyta (Brown Algae)

- Phylum Myxomycota
- Phylum Oomycota

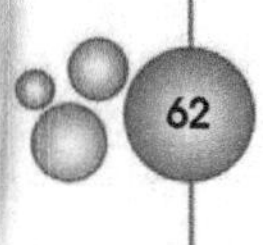

A. ANIMAL LIKE PROTISTS(PROTOZOA)

includes protists that in Since they are more similar to other protists than to animals, they are placed in this group.

1.Rhizopoda (Root-footed protozoa)

Rhizopoda means "root-footed" in Latin (Rhizo=root, poda=feet). These are unicellular organisms living in soil, seas and fresh water. They are harmless in general although some are parasitic. cytoplasmic extensions called Amoeba and Radiolaria are examples of this group.

a.

Amoeba can survive dry conditions by forming a protective covering called a cyst. They are found in ponds and streams. Reproduction is asexual through binary fission.

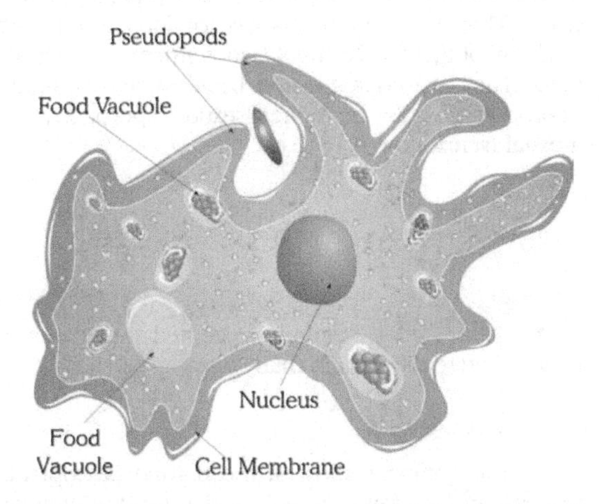

Figure 8.1.: Amoeba Structure. Since amoeba are enclosed with a stiff layer around the plasma membrane, they don't have a fixed shape.

b.

The causative agents are cysts transported via unclean hands and food. After opening, the cysts turn into amoeboid cells and infect the inside of the digestive system. The organism lives as a parasite of the large intestine where they destroy the cells coating the inner wall,
Examination of feces is necessary for detection of cysts with 4 nuclei.

c. Radiolaria

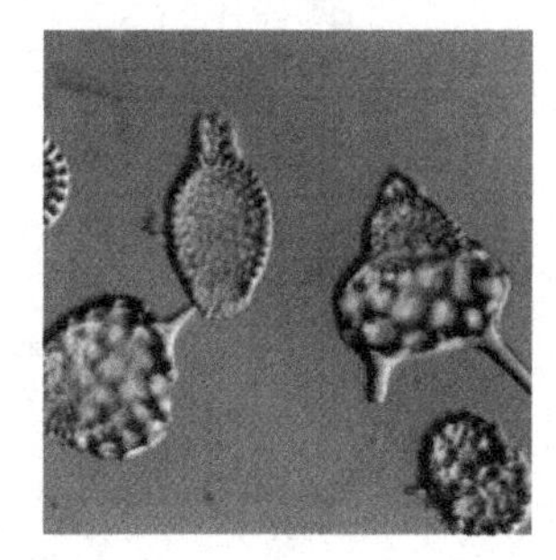

These are marine organisms found near the surface. They have thinner, filament-shaped pseudopods and an exoskeleton made of SiO_2. Since they live in the sea, they do not possess contractile vacuoles.

2. Foraminifera

They are found in warmer and deeper parts of the sea. The soft part of their bodies is covered by an exoskeleton made of $CaCO_3$. As they die, they sink to the ocean floor to form chalk layers in geological periods. One gram of sea sand may contain 50,000 foraminifera.

3. Actinopoda (Ray animalcules)

These ball-shaped marine organisms are covered with filamentous cytoplasmic projections called axopods. They are about as big as a pinhead. After death, their remnants form a solid layer on the ocean floor which is later petrified.

4. Zoomastigina

Anteriorly these have flagella which act in agile locomotion and food capture. As they lack chloroplasts, they are mostly free heterotrophs, but some are also pathogens eg. Trypanosoma gambiense and Leishmania tropica.

a.

(trypanosomiasis), (Glossina palpalis). The parasites are transferred by the bite of the fly.

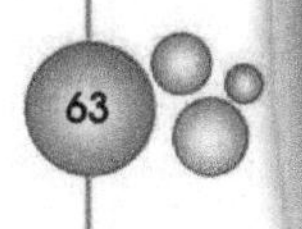

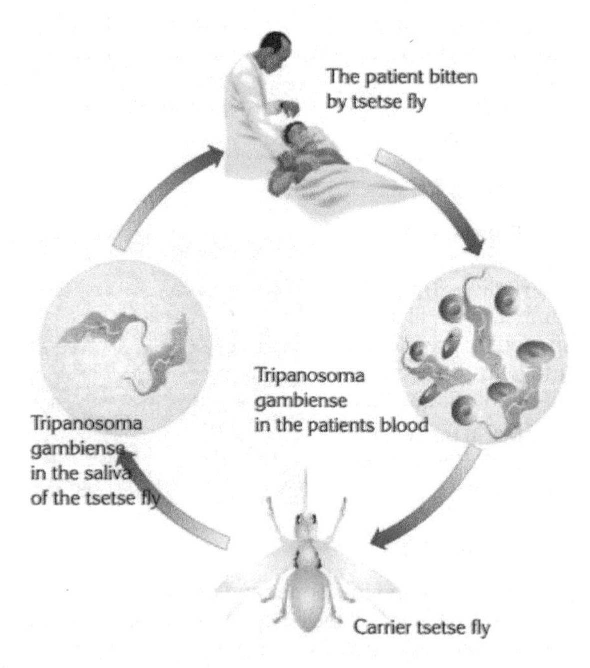

Figure 8.2.: The life cycle of Trypanosoma gambiense, which causes sleeping sickness in humans

b. Leishmania tropica

This parasite causes leishmaniasis on the skin of the arms, legs, hands etc.

5. Ciliophora (Ciliates)

[illegible] is the most advanced group of unicellular [illegible]. The hard outer part of the cell, called the pellicle, [illegible] trichocysts that are discharged during defense and predation. There are [illegible]: the macronucleus (used in metabolism) and the micronucleus (used in reproduction). There is also a cytostome (mouth) which opens to the cytopharynx (cell pharynx), a cytostome (cell anus), vacuoles, [illegible] The last provides osmoregulation within the cell by discharging excess water. Waste products are removed through excretory vacuoles. Reproduction is usually asexual by binary fission or conjugation. The latter contributes genetic diversity to the population. [illegible] Vorticella, and Stentor are [illegible]s [illegible] (Figure 8.3).

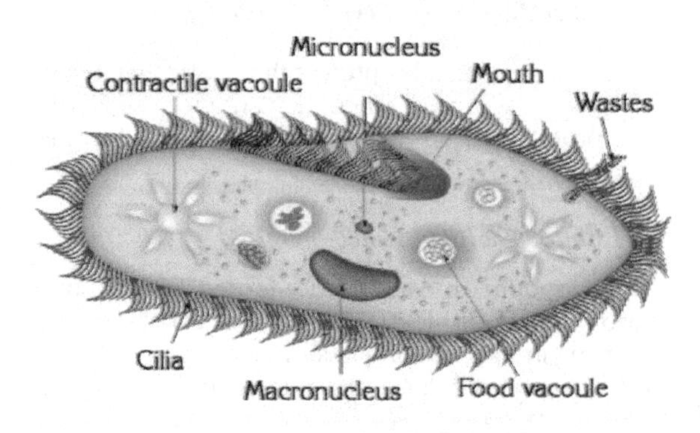

Figure 8.3.:Paramecium. Since the cell surface isn't flexible, paramecia have a definite shape. They have parts that function like the mouth, anus etc. of higher organisms.

6. Apicomplexa

All species are endoparasites of vertebrates and invertebrates. Because of their parasitic lifestyle they do not possess contractile vacuoles or phagocytic vacuoles. They have no organelles for locomotion, but flexibility of their body enables movement via contractions. They are characterized by an apical complex. Reproduction can be either sexual (sporogony) or asexual (schizogony).

B. [illegible]

[illegible] but aren't accepted as plants since they are quite different in many ways.

1. Dinoflagellata

These comprise most of the marine phytoplankton. Except for a few colonial species, most are unicellular. [illegible] pigmented (chlorophyll a and b, carotenoids), [illegible] protists with two flagella at right angles to each other and a shell made of cellulose. Those lacking pigmentation feed on microorganisms. Some dinoflagellates without flagella or shell live inside jellyfish, corals, and mollusks endosymbiotically. Photosynthetic, commensal, and parasitic species exist.

MALARIA

The vector of this parasite is the Anopheles mosquito (definitive host), while the intermediate

and epithelial cells of humans.

etc. The female Anopheles mosquito sucks blood from warm-blooded animals and transmits malaria parasites from one to another. The males do not transmit the disease as they feed only on plant juices.

and the liver becomes tender which may lead to cirrhosis. Additionally, swelling of the belly and anemia are observed, while occasionally shock and bleeding occur when capillaries are blocked by diseased erythrocytes.

Typical malaria attacks

1. Cold (shaking) stage (1/2-2 hours): There is a consistent shivering and feeling of cold. Headache and nausea may be present.
2. Fever stage (ca. 24 hours): The fever may be as high as 40-41 °C, accompanied by nausea, vomiting, headache, and cold sore on the lips.
3. Wet stage: Body temperature, spleen size, and sleep all become normal, and sweating begins. The normal period continues until the next cycle of paroxysms begins.

Malaria is also seen in reptiles, birds, and other mammals besides humans.

Each Plasmodium species causes different types of malaria, for example:

P. falciparum: *Plasmodium falciparum* causes malaria quartana. The schizogony period inside the erythrocytes lasts 72 hours.

P. vivax: *Plasmodium vivax* causes malaria tertiana. The schizogony period is around 48 hours.

P. malaria: *Plasmodium malaria* causes malign malaria tertiana (tropical malaria). The schizogony period is ca. 24-36 hours (*Figure 8.4*).

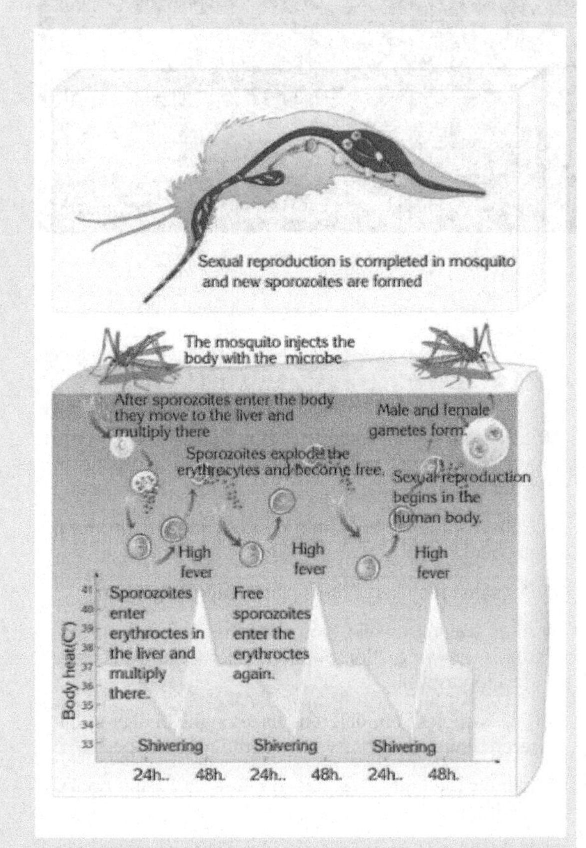

Figure 8.4.: Shaking stages and temperature changes in the life cycle of malaria-causing Plasmodium in humans.

2. Bacillariophyta (Diatoms)

Most are unicellular except for some which are colonial. Radially or bilaterally symmetric, they are mostly free, or occasionally attached, plankton species of fresh waters and cold seas. They move by gliding on a slimy matter produced by the shell. Photosynthetic pigments like chlorophyll a and b, and carotenoids give a yellowish or brownish color. They store food in the form of carbohydrates or lipids. With many species throughout many water ecosystems they constitute the most important group of primary producers (Figure 8.5).

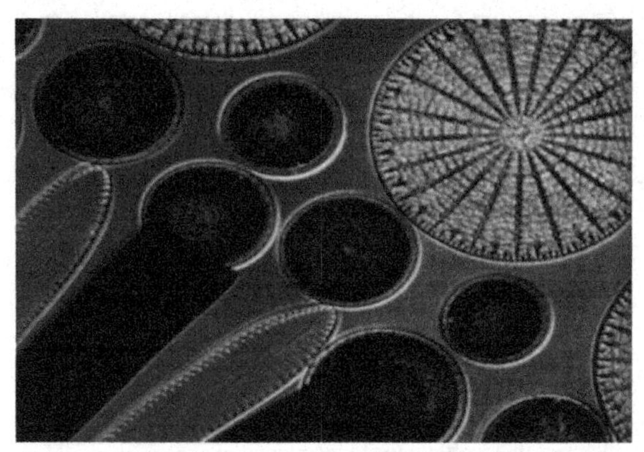

Figure 8.5.: Diatoms of various shapes are responsible for most of the production in water ecosystems.

PROTOZOA CULTURE

Most protists are microscopic. It is not difficult to observe these cells. The procedure for observing these organisms is given below.

- Collect some water from a pond or small river into a big beaker.
- Put some remains of dead plants into the beaker.
- After waiting awhile, you will observe some protozoans growing. Some other microorganisms may develop as well.
- Take samples from different places in the beaker at different times and observe under the microscope.
- Identify the organisms that you observed using books, the Internet etc.
- Make a list of organisms you identified and record the changes over time.

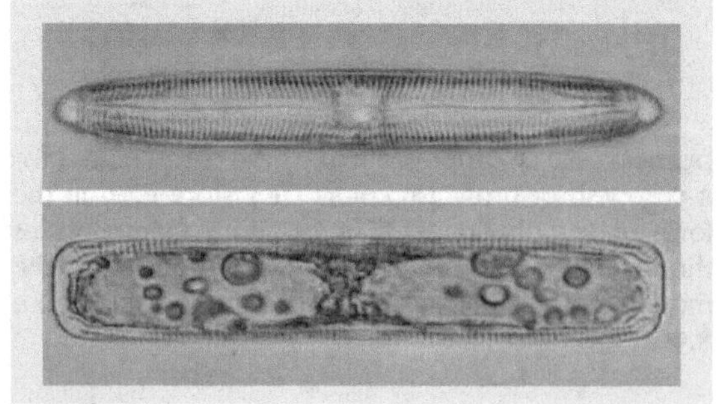

3. Euglenophyta

One-third of this unicellular group contains photosynthetic pigments. The heterotrophs lack the pigment and are colorless. Euglenophytes can be found in fresh water rich in organic content. Generally they possess 2 (one shorter) flagella, and the euglena is representative of the group.

Euglena

unicellular organisms and are photosynthetic in the presence of light. . Carbohydrates are stored as paramylon, a type of polysaccharide. The organism shows both animal and plant characteristics.

Locomotion is made possible by flagella situated at one end of the cell. Next to the flagella there is a contractile vacuole and stigma (eye spot). The pellicle, a protective layer that gives a definite shape to the cell, covers the outer surface (Figure 8.6).

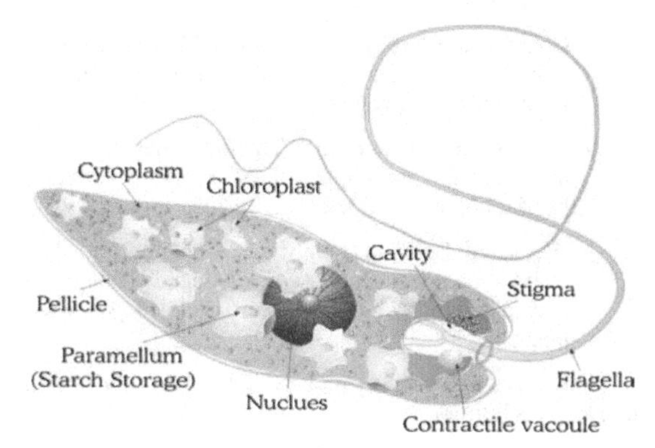

Figure 8.6.: Euglena, an organism that is both heterotroph and autotroph

mostly and reproduction is asexual by longitudinal cell division (Figure 8.7).

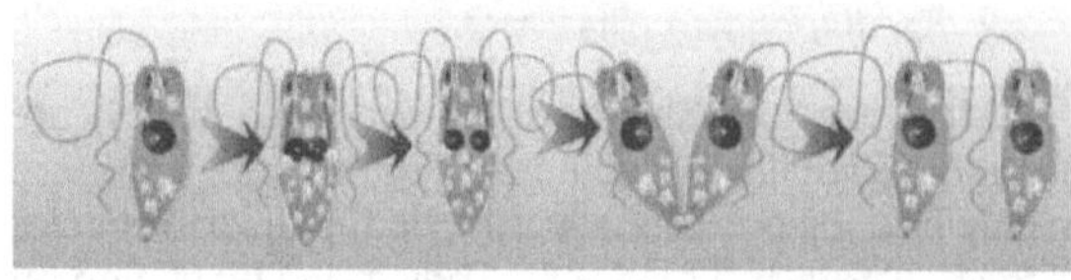

Figure 8.7.: Euglena reproduces asexually by longitudinal cell division.

4. Chlorophyta (Green Algae)

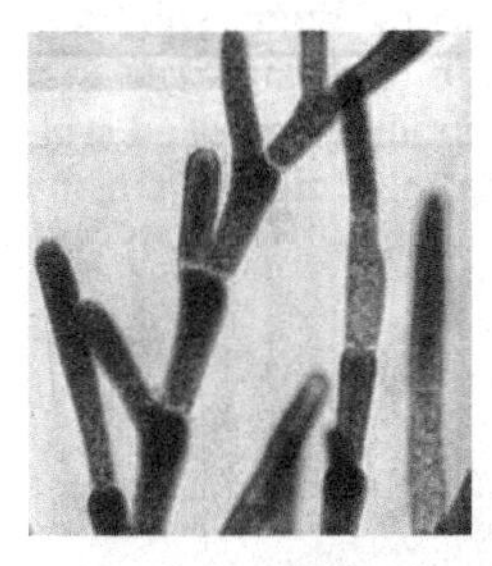

These organisms are unicellular or colonial species, as well as multicellular species that are not organized into tissue. Reproduction may be sexual or asexual. Although many have flagella at least in a certain period of their lives, some species are fully sessile. Most species live in freshwater and there are many species of significant ecologic and economic importance. They carry photosynthetic pigments like chlorophyll, carotene, and xanthophylls. They store carbohydrates as starch in chloroplasts. The cell walls are made of cellulose, which is why algae are used in the paper industry. (Figure 8.8).

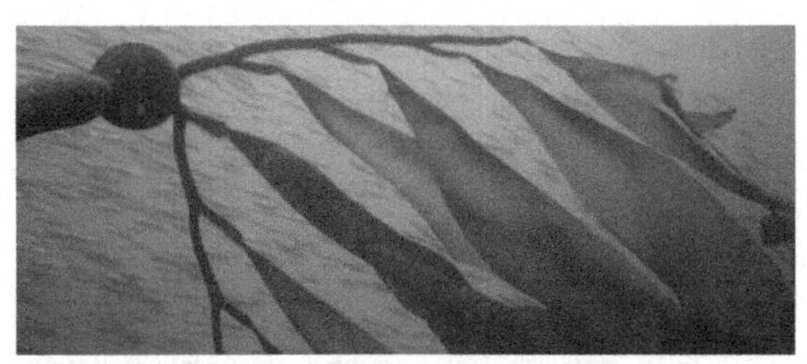

Figure 8.8.: *Green algae used in paper and food production are cultivated commercially.*

Sea lettuce (Ulva lactuca), Spirogyra,

a. Spirogyra

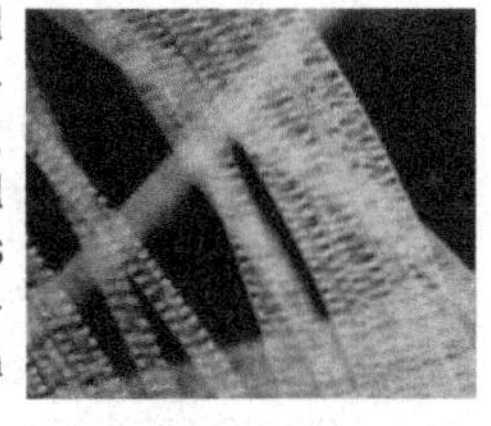

These are filament-shaped organisms found on the surface of fresh water. In autumn, Spirogyra produces a bad smell (like dead fish) in bodies of water. They are also responsible for the foaming of fresh water.

b. Ulva lactuca (Sea Lettuce)

Sea lettuce looks like creased sheets of green cellophane, 30 cm in length and with stalk-like rhizoids for attachment to the substrate. They are particularly found in saltwater. Being rich in nitrogen, they are used as fertilizer. Rich in carbohydrates and vitamins A, B and C, they are used as food in Japan and China. During cell division, the nuclear membrane doesn't disappear. Metagenesis (alternation of generations) is observed in the organism: diploid sporophytes produce haploid spores that will form gametophytes.

c. Chlamydomonas

These are unicellular organisms with two vacuoles, two flagella, one stigma, one nucleus and a cup-shaped chloroplast. Reproduction is asexual, or sexual with isogamy. Some unicellular algae don't separate when dividing and remain together in a gel-like substance. This is called a colony. Colonies of green algae may be formed by sets of 4, 8, or 16 cells.

Some biologists accept these as transitional between unicellular and multicellular organisms. Advanced features of multicellular organisms not found in colonial protists include the connections between cells, and the organization of cells in structure, shape, and function.

d. Pandorina

This freshwater organism forms a colony of 16 cells covered by a 2-layered gelatinous envelope. All cells are identical in function and morphology. There is no hierarchy or cooperation between cells. When a cell leaves the colony, it forms a new 16-cell colony. The cells' flagella work in harmony to move and position the colony. This is the simplest type of social organization.

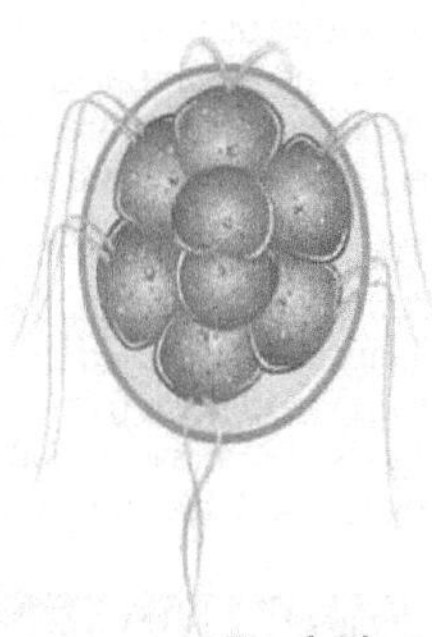

Pandorina

e. Eudorina

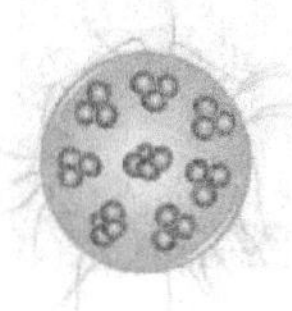

Eudorina colonies consist of 32 cells and are compressed laterally. Therefore, this is the simplest colonial organism, having both "posterior" and "anterior" sides.

ECONOMIC IMPORTANCE OF ALGAE

Use of algae as food dates back to 2700 B.C. in island cultures with a limited variety of food sources. Today, there are some new, economically beneficial practices involving aquatic plants and algae.

Spirulina, a non-toxic blue-green alga, is cultivated and used as a food. Also some nitrogen-fixing blue-green algae are used as fertilizers. Diatom shells, containing silica, accumulate in geological layers. This "diatomaceous earth" was formerly used to soak up nitroglycerine in dynamite production. Today it is used in electrode bars and flame-resistant bricks.

Some microscopic green algae (Cladophora and Chaetomorpha) whose cell walls are made of cellulose are used in the paper industry. This is being done with positive results in Turkey. Furthermore, sea lettuce (*Ulva lactuca*) is eaten as a salad and used as a tea in Japan

Red and brown algae, found mostly deep in the sea, are economically important in many ways. Algae are sources for iodine and bromine. Brown algae, especially Macrocystis porifera, which is abundant and cultivated in seas, as well Laminaria, Cystoceria, Sargassum, and Ascophyllum sp. are used commonly in many industries, eg. textile, food, and dye.

The Porifera genus of red algae is cultivated and consumed in Japan. Another product obtained from red algae is agar. Agar is used in the food industry for production of some food items (candies, ice-cream, mayonnaise, canned foods), but it is especially used in preparing bacteriological media and in dentistry.

f. Volvox

Members of this genus form huge spherical colonies consisting of thousands of cells arranged in a single outer layer enclosing an inner space filled with mucilage. All cells have cytoplasmic bridges (plasma bridges).

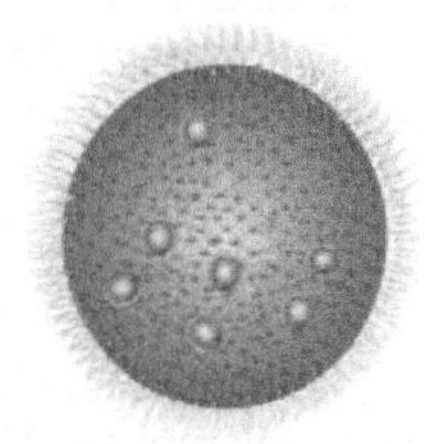

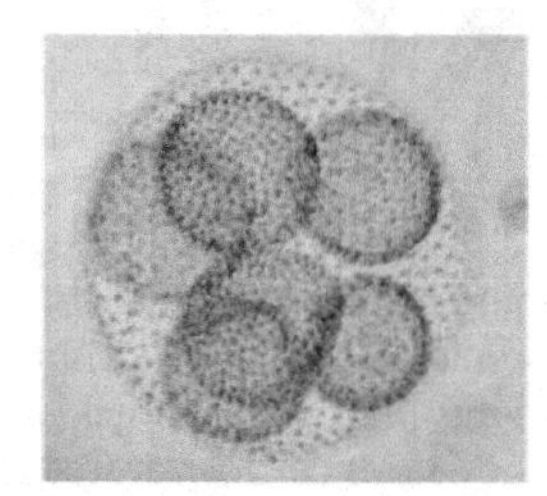

Figure 8.9.:*Volvox colony and a new volvox colony formed by asexual reproduction.*

There are 3 types of cells in Volvox colonies:

- Outer somatic cells: These are small cells with stigma, contractile vacuole, chloroplasts, and two flagella. Their function is in locomotion and feeding.
- Inner somatic cells: These have contractile vacuoles and stigma, but no flagellum. They provide osmotic balance and removal of wastes from the colony.
- Germ cells: These larger cells are found inside the colony.

This type of differentiation (germ and somatic cells) is seen first in Volvox. There is asexual and sexual reproduction. The first is by formation of a daughter colony, while for the second haploid cells (sperm, egg) must be produced.

As seen above, there is specialization of Volvox cells for feeding, locomotion, and reproduction.

Importance of specialization

- Specialized structures and organization provide economical and productive use of energy and speed up metabolic activities.
- Compared to minute unicellular animals, larger multicellular (and colonial) organisms are less threatened by predation and other factors.

COMPARISON OF ORGANISMS THAT FORM THE KINGDOM PROTISTA				
PHYLUM	**MORPHOLOGY**	**MOVEMENT**	**PHOTOSYNTHETIC PIGMENTS**	**SOME OTHER FEATURES**
Rhizopoda (with false feet)	Single celled protists with shells; no body shape	With false feet	No	Some have shells
Foraminifera (Porous)	Single celled	With cytoplasmic extensions	No	Have a porous shell
Actinopoda (sun animals)	Single celled	Some have flagellated sex cells	No	Extensions called axopods arising out of porous skeleton
Zoomastigina (Animal-like flagellates)	Single celled	Move with one or more flagella; some are ameoboid	No	The symbiotic species are well-specialized
Ciliophora (Ciliates)	Single celled	Move with cilia	No	Have a micro- and a macronucleus
Apicomplexa (Sporozoa)	Single celled	Nonmotile	No	Most are parasites and form strong spores
Dinoflagellata (plant-like flagellates)	Single celled, some form colonies	Have two flagella	Chlorophyll a and c; Caretenoid; Fucoxanthin	Most have a cover made of cellulose
Bacillariophyta (Diatoms)	Single celled, some form colonies	Most are sessile; some move by sliding	Chlorophyll a and c; Caretenoid; Fucoxanthin	Have a shell made of silica
Euglenophyta	Single celled	One long and one short flagella	Chlorophyll a and c; Caretenoid	A flexible cover is present around the body
Chlorophyta (GreenAlgae)	Single celled	Most have a flagella in some part of their life and not in others	Chlorophyll a and c; Caretenoid; Fucoxanthin	Various types of reproduction
Rhodophyta (Red algae)	Most are multicellular; some are single celled	Sessile	Chlorophyll a and c; Caretenoid; phycocyanin; Phycoerythrin	Form a shallow layer
Phaeophyta (Black algae)	multicellular	Have a pair of flagella in sex cells	Chlorophyll a and c; Caretenoid; phucocyanin	Leaf and branches form on body
Myxomycota (Slime molds)	Multinucleated cells	Fluid cytoplasm; flagellated or amoeboid movement	No	Reproduction with spores in spore sacs
Oomycota (Water Molds)	Micellium	Zoospores with two flagella	No	Cell walls made of cellulose and/or chitin

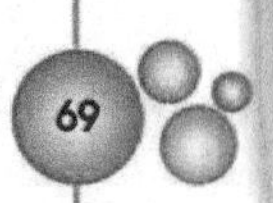

- Multicellular organisms are more resistant to adverse conditions. They can also survive the loss of a few cells.

Specialization of Complex Organisms

A unicellular organism has to perform all basic metabolic activities. But a specialized cell does one particular work. In other words:

- Food is transferred to all cells via a circulatory system.
- Wastes are removed by an excretory system.
- Reproduction is performed by a genital system.
- Coordination within and between cells is provided by controlling systems.
- The more specialized the body part, the more it is modified. Thus specialized organs are more vulnerable.
- Injury or damage to a tissue affects the whole organism.

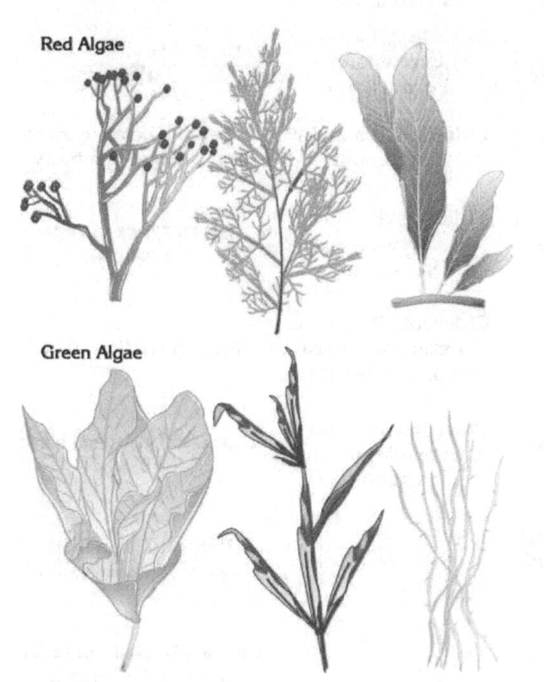

Figure 8.10.: Algae types. Algae are essential for sea life and humans.

5. Rhodophyta (Red algae)

(>100 m). Except for a few unicellular examples, they are multicellular (Figure 8.10).

Their photosynthetic pigments are chlorophyll and red pigments (phycoerythrine). These absorb blue light and UV (phycoerythrine). Phycoerythrine absorbs light of shorter wavelengths, which enables life in deeper seas where other algae are absent. Storage molecule is a special starch called floridean. There is no motile cell in any part of the life cycle. Stem is densely packed with filaments lying in the gelatinous matrix.

Figure 8.11.: Detailed structure of a brown alga. Existence of life in seas and oceans partially depends on algae. Their roles as oxygen producers and as a food source are important.

6.

varying size (ca. 1 cm to 60 m) (Figure 8.11-12). The

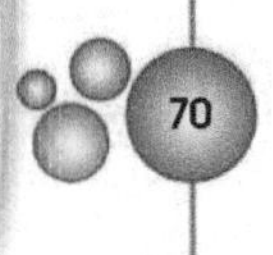

photosynthetic pigments are chlorophyll and fucoxanthine. Fucoxanthine, which is abundant in cells, covers the chlorophyll and gives the cell a yellowish to dark brown or black color. With the help of this pigment brown algae can live in deep waters. Laminarine is the storage molecule of brown algae.

Motile cells are biflagellate. Algae attach themselves to the ground with rhizoids. Besides nutritive use, I, K, P are obtained from brown algae (like Laminaria, Sargassum, Fucus). Fibers resembling artificial silk are used in industry.

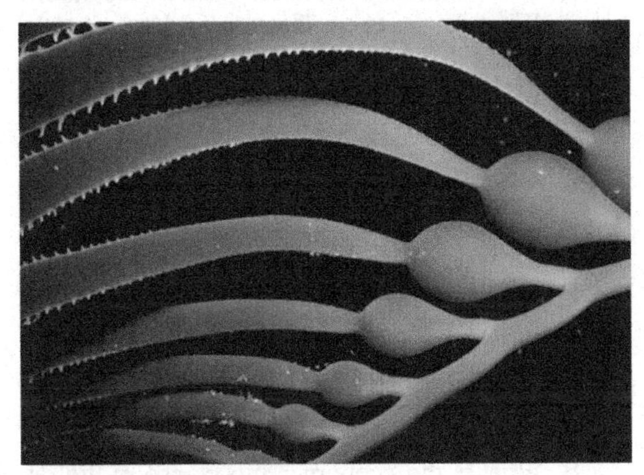

Figure 8.12.:Algae is important to sea life and to human life.

C.

, as they aren't photosynthetic, they are formed by filamentous hyphae. But unlike fungi most of these have flagella, centrioles and

1. Myxomycota

Uncertainty about the systematic position of slime molds results from some unique features of the group. They are amorphous, multinuclear, saprophytic or seldom parasitic organisms with amoeboid movement. They don't have cell walls covering the plasma. Therefore they are cytoplasmic masses. Food capture is done by amoeboid movement. Moist forests, tree stems, and decaying leaves are habitats of these organisms. Storage molecule is glycogen. Some are parasites of plants, causing tuberous roots in cabbage. There are more than 550 species. Reproduction is sexual or asexual. Sexual reproduction is made by two zoospores, whereas asexual reproduction is made by thick-coated spores.

2. Oomycota

In this group there are over 5000 aquatic or terrestrial species. recognized as Because of their considerable resemblance to fungi, they were long classified under the latter. They develop hyphae, then mycelia as in the fungi. Cell wall may be chitinous (like fungi), cellulous (as in plants), or both. In certain terms of their lives they carry flagella, which is the reason for their present position in protista. In good conditions they reproduce asexually, in worse conditions they reproduce sexually with oospores formed by oogonia.

OOMYCOTA

(Approxiamately 500 spp)

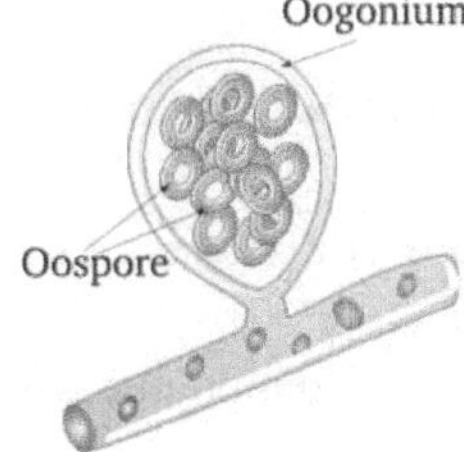

There are aquatic or terrestrial species. Sexual reproduction is made by oospores produced by oogonia, while asexual reproduction is by flagellate spores. Diploid hyphae have chitinous or cellulous cell walls. They form white spots an garbage and plants.

SELF CHECK

PROTISTS

A. Key Terms

Rhizopoda	Pseudopodia
Foraminifera	Euglenophyta
Actinopoda	Chlorophyta
Zoomastigana	Rhadophyta
Ciliophora	Phaeophyta
Apicomplexa	Myxomycota
Dinoflagellata	Oomycota
Bacillariophyta	Colony
Mutualism	Malaria

B. Review Questions

1. What are the differences between bacteria and protists?
2. How do algae obtain food?
3. Describe the size and color of algae in the six different groups.
4. List the products that are made with substances from algae.
5. Compare the cell structure of a euglena to a flagellate protozoan.
6. Write the names of the most common diseases caused by protists.
7. What are the main differences between Dinoflagellata, Ciliophora and Myxomycota?
8. What are the features of a typical protist? Why are the prostists so difficult to characterize?
9. How are the protists important to humans?
10. How are they important ecologically?
11. Kelps are brown algae that grow in the ocean. What is the main importance of kelps in marine habitats?
12. Why does a euglena need an eyespot?
13. How does a paramecium get its food?
14. How does an amoeba reproduce?
15. What do all protists have in common?
16. Compare how amoeba, euglena and paramecium move.
17. How are protists different from bacteria?
18. What are algae, functionally and taxonomically? Why do most not consider them to be plants?
19. What is unusual about the cell wall of diatoms? What happens as diatoms undergo asexual reproduction?
20. Write the similarities and differences between plants and algae.
21. What are the differences between flagella and cilia?
22. What was the calamity caused by water molds in Ireland in 1845-1846?
23. What are red tides? Which organims cause them?
24. Why do algae grow mostly near the surface of water?

25\. What is the economic importance of the different types of algae?

25\. What is a kelp, and what enables it to float on the water?

26\. Distinguish among each of the following protozoan groups:

(a) amoebas, foraminiferans, and actinopods

(b) flagellates, ciliates, and sporozoans.

C. True or False

1. ☐ Malaria is caused by plasmodium.

2. ☐ Entamoeba histolytica is pathogenic protists.

3. ☐ Animal like protists are non-plotosynthetic.

4. ☐ Plant like protists are motile.

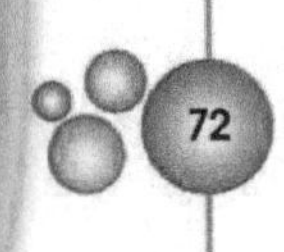

D. Matching

a. Parasite	( ) root-footed
b. Mutualism	( ) false-footed
c. Rhizopoda	( ) feed on other organisms
d. Pseudopodia	( ) cooperate to survive
e. Sea lettuce	( ) Ulva lactuca
f. Rhodophyta	( ) red algae

E. Fill in the blank

1. Protists that ingest their food as animals do are informally called ....
2. Amoebas move and obtain their food by means of ....
3. Paramecium and other .... move by cilia.
4. The .... are a group of parasitic protozoa that form spores at some stage in their lives.
5. A dinoflagellate bloom is known as a ....
6. Chlorophyll a, chlorophyll b , and carotenoids are found in green algae, ..., and plants.

F. Multiple choice

1. The amoeba moves by means of........................

A) Peristaltis B) Pinocytosis C) Cilia D) Flagellata E) Pseudopodia

2. Which structures are used for locomotion in some protists?

A) Tentacles B) Spore C) Cilia D) Smoot muscles E) All of them

3. Which of the following <u>is not</u> property of protista?

A) They are unicellular organisms

B) They are eucaryotic cells

C) They have flagella

D) Some members of protista are parasitic

E) They reproduce by cell division

4. Trypanosoma is example of...................

A) Flagellata B) Sporozoa C) Ciliata D) Rhizopoda E) Vibrio

5. Which of the following organisms like both plants and animals?

A) Paramecium B) Euglena C) Plasmodium D) Trypanosoma E) Amoeba

6. Which of the following organisim has chloroplast?

A) Paramecium B) Euglena C) Plasmodium D) Trypanosoma E) Amoeba

IX. KINGDOM FUNGI

A.

organisms which are filamentous or rarely unicellular. They are , but there are many aquatic (or marine) species. higher In all stages of the life cycle, cells lack flagella or cilia. All fungi have chitin materials in their cell walls. Although there are some similarities with algae, they are separate from all similar groups because they lack chlorophyll. exclusively

All except for yeasts have hyphal structures. Hyphae are colorless, slender, long filaments forming an interwoven mass called a mycelium. All the mycelia form a thallus, the body of the mushroom. All parts of a mushroom thallus are homogenous. Mushrooms feed and reproduce vegetatively by means of the mycelia. Food is not stored in the form of starch, but rather as lipids and glycogen.

Reproductive cycles include both sexual and asexual phases. Most are haploid, and only the zygote is diploid. Some animal or plant parasites cause disease. Some types of yeasts are used in the fermentation of vinegar and alcohol.

1.

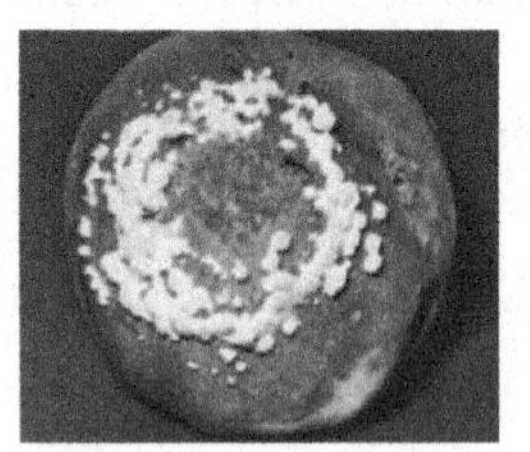

The mycelia of molds are well-developed, hyphae are cellulose and rarely divided by septa. Like algae, molds reproduce asexually with spores, and sexually with gametes. Since septa are quite rare, the cytoplasm of the hyphae bear many nuclei.

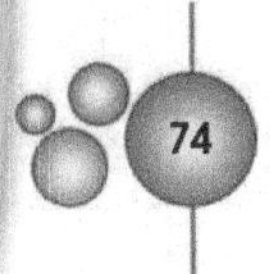

a. Candida albicans

is among the most abundant pathogenic fungi in nature. Quick-flourishing mycelia , mucosa, internal organs and beneath the . Thrush is seen in oral mucosa of babies, and in nasal mucosa and retina of adults. Furthermore, the fungus may infect finger- and toenails (cnychomycosis) resulting in pain and inflammation, and finally thickening, crumbling, and darkening of the nail.

b. Or bread mold.

decaying organic matter and is found especially on Rhizoids penetrate the substrate (bread) and other hyphae (stolons) grow vertically as long stalks holding spherical sporangia. Sporangia bear many black spores.

c. Rhizopus stolonifer (Black bread mold)

Bread Mold Culture

You may encounter many mold species in daily life. One of these, bread mold, can be observed in the laboratory with the following procedure:

- Moisten a slice of bread or a tomato and place it in a petri dish (or any glass container with a cover).
- Collect some dust on your finger and spread it around on the slice. You don't need to wory about the presence of spores--they are found almost everywhere.
- Close the cover of the petri dish..
- A few days later cotton-like structures (mycelia) should appear, followed by black spots.
- Take a sample from the culture and put it on a slide with a few drops of water.
- Draw what you see under microscope and discuss with your friends.

Figure 9.1.: The life cycle of bread mold. This fungus forms a black cover on bread and similar foods. Reproductive cycle includes asexual (spores) and sexual (isogamy) phases.

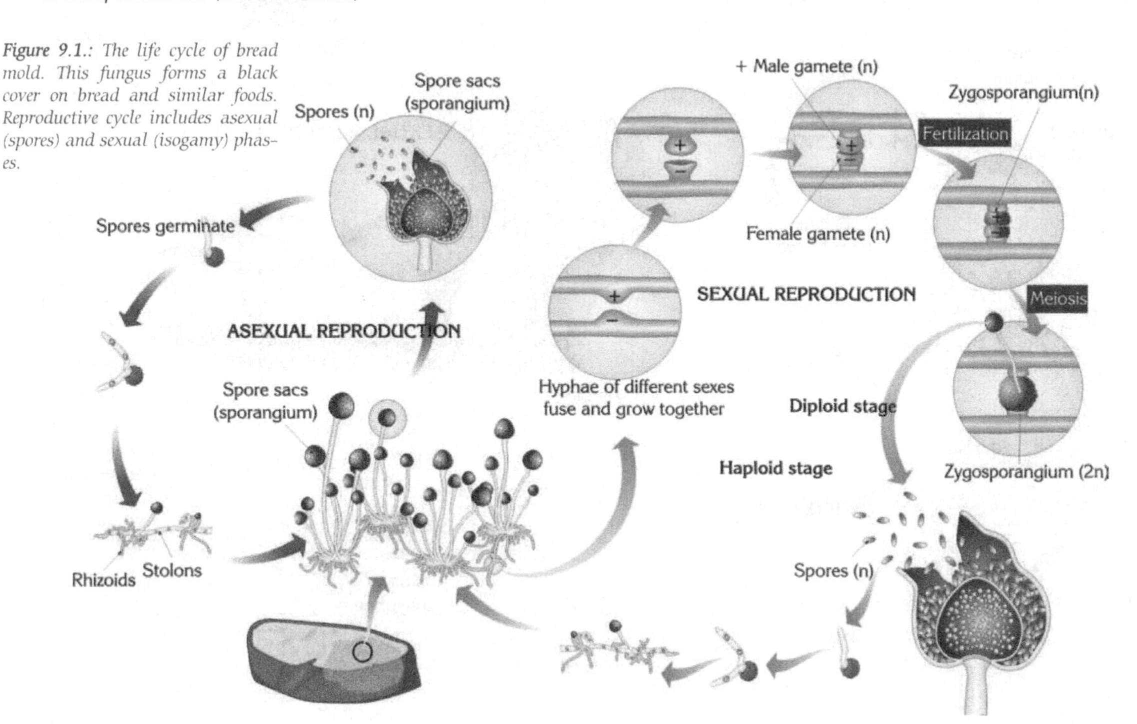

Rhiozopus stolonifer (black bread mold) lives on some foods (bread, etc.). Spores dropped on the bread form hyphae. These get nutrients from the bread. Sporangia are formed at the tip of stalks. Both sexual and asexual reproduction is observed. (*Figure-9.1*).

2. Ascomycota (Ascomycetes, Sac Fungi)

including Nevrospora molds. Hyphae have septa, but the cross walls are perforated. Sexual germ cells (ascopores) are produced by pouch-like cells called asci (sing. ascus). Spores are released by squirting. Many sac fungi hyphae form a cup shaped structure (*Figure 9.2*). Asexual reproduction in many species involves production of spores, conidia (sing. conidium), in conidiophores. Examples are yeasts and symbiont fungi in lichens.

a. Saccharomyces cerevisiae (Yeast)

These do not have hyphae and reproduce by budding and rarely via ascospores. , leavening bread. the sponge-like

b. Penicillium

These live saprophytically on cheese, lemons and some other foods, creating blueish spots. They resemble a brush in shape. Some species are responsible for the flavor of Roquefort, Camembert and Gorgonzola cheeses. Formerly, cheese containing penicillium was given to tuberculosis patients. Penicillium produces penicillin, the antibiotic. , and carbuncle. Some patients are allergic to it, so use without a doctor's supervision or in unnecesarry cases may be dangerous.

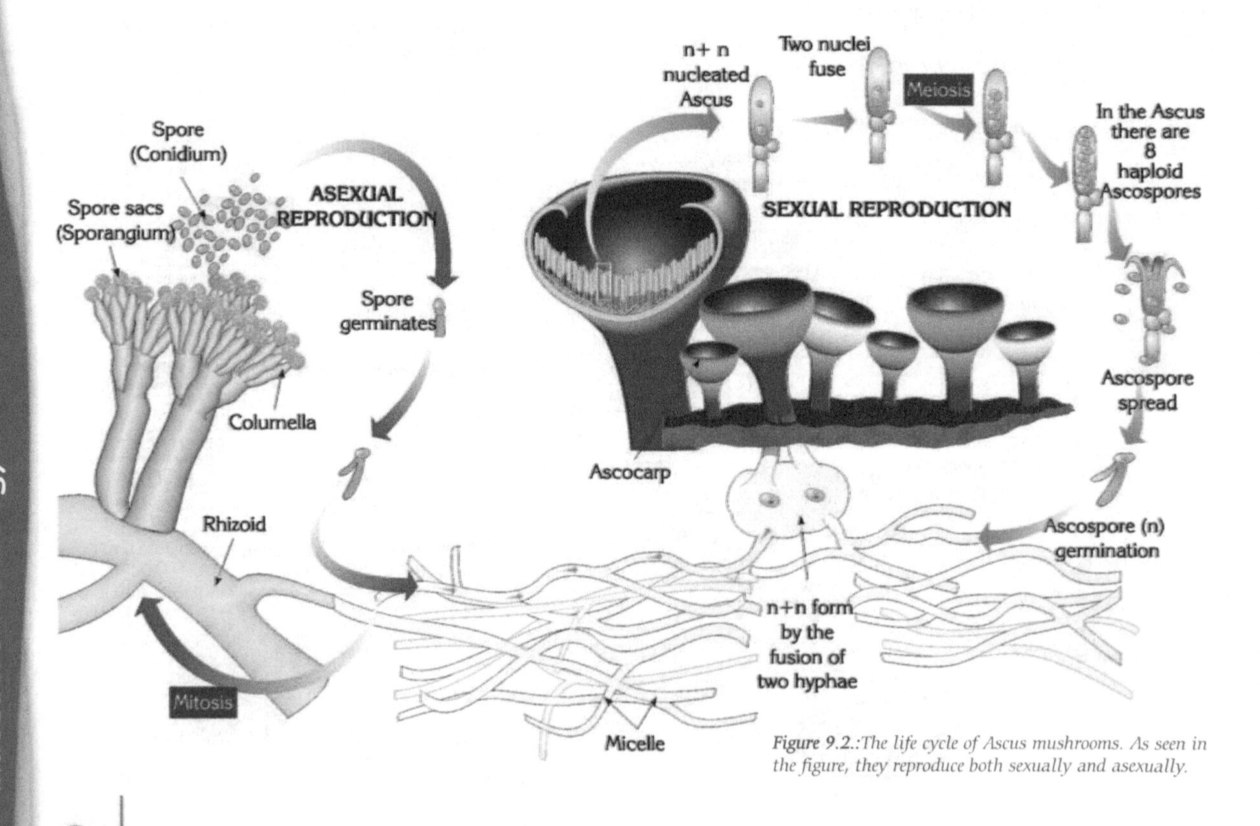

Figure 9.2.:The life cycle of Ascus mushrooms. As seen in the figure, they reproduce both sexually and asexually.

Asexual spores are conidia which develop inside conidiophores.

3. Basidiomycota (Club Fungi)

, and harmless terrestrial Hyphae have septa. The club-shaped, aboveground part of the mushroom is called a basidiocarp. Under the umbrella-shaped part of the basidiocarp, basidia develop in abundance inside the lamellae. Mature basidiospores are the means of asexual reproduction (*Figure 9.3*).

Only mushrooms that are definitely known to be edible should be eaten. Sometimes it is very difficult to differentiate poisonous species, which even kill humans. The symptoms usually appear after 8 hours. By the end of this time the liver and kidneys fail and death is inevitable. The only hope is kidney or liver transplantation. Consequently, an expert eye is needed to identify edible species which, even so, may be confusing in some cases.

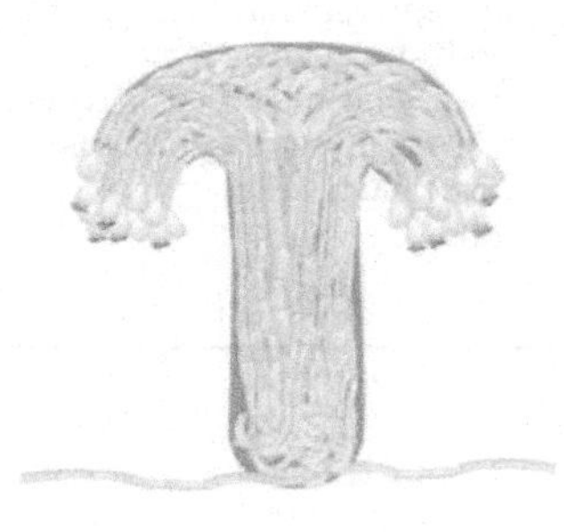

Figure 9.3.: Structures in mushrooms. Mycelium units form the thallus structure. All parts of the thallus are homogenous. Additionally mycelia are the means of nutrition and reproduction.

4.

like ringworm, Since sexual reproduction isn't observed, Many seem to be related to sac fungi as they both reproduce by spores, but several species are similar to basidiomycetes.

B.

Consequently, the systematic position of lichens is unclear. Because of the physiological and morphological association between the two, lichens show both fungal and algal characteristics. Fungi in lichens are sac or club fungi, algae are blue-green or green algae. External shape is determined by the fungal component.

In the relationship, the in return. This is mutualism. Lichens are autotrophs, despite their fungal component, and are found in even the most inhospitable environments (mostly in rocks and trees as ink spots).

They flourish in clean air, but they do not develop in areas with polluted air. They can therefore be used as an indicator for air pollution in an area.

According to their external features, lichens are divided into 3 groups:

- Shell-like lichens: Thallus is flattened like a shell. Lichen is firmly attached to the substrate (eg.leconora).
- Leaf-like lichens: Thallus is formed by lobes of different sizes and is loosely attached to the substrate (eg.Parmelia).
- Branch-like lichens: Thallus may be either vertical or hanging down from the trees, filamentous or sheet-like in shape. They are mostly branched and only a small part is attached to the substrate (eg.usnea).

The algae in lichens reproduce vegetatively. The active role in reproduction belongs to the fungal part. When the fungal spores are released they sprout and, upon contact with a suitable algae, the symbiosis begins again. Products of lichen include lichen acids and arsein (stain).

ECOLOGICAL AND ECONOMIC

Since many fungi live as saprophytes they have an important role in ecology. They absorb nutrients from organic matter and dead organisms. During this process water, CO_2, and minerals are released back into the environment. The absence of this work would destroy an ecosystem.

Some fungi, like lichens and mycorrhizae, have mutualistic life-styles. Mycorrhizae is an association between specific fungi and plant roots. In this symbiotic relationship the plant gets water and minerals from the fungus, and in return the fungus obtains nutrients (glucose, amino acids, etc.) from the plant. This relationship is essential especially during the growth period for many plants. Mycorrhizae are highly susceptible to acid rain and this is becoming another threat to boreal forests.

Fungi have great economic importance. Nearly 200 edible species are cultivated and consumed in large amounts. On the other hand, many deaths and poisonings are caused by wild mushrooms. Fungi include many plant and animal pathogens as well.

Apart from food, fungi are used in the baking of bread, the fermentation of alcohol, and the production of various antibiotics. Pain-killing drugs are also obtained from fungi. Some fungi are used in the production of citric acid and other chemicals. Recently, hormone production from fungi has been started using recombinant DNA technology.

HMW: DRAW TABLE

FUNGI AND THEIR FEATURES

Fungi are organisms with cell walls and hyphae. They are mostly terrestrial and saprophytic or parasitic organisms.Their reproduction can be sexual or asexual (via spores)

FUNGI	FEATURE
ZYGOMYCOTA (includes nearly 800 species) 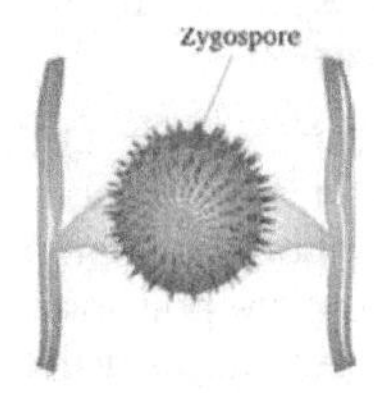	Sexual reproduction occurs via zygospores. The cell wall is made up of chitin. An example is bread mold.
ASCOMYCOTA (includes nearly 30,000 species) 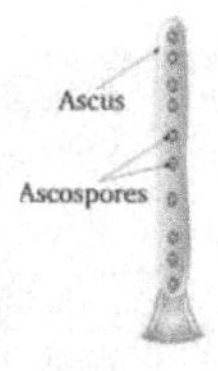	They are also called sacked mushrooms. Ascospores formed in these sacs enable sexual reproduction. There are both unicelullar types, eg. yeast, and multicellular types with septa, eg. Aspergillus and Penicillium, that grows on food.
BASIDIOMYCOTA (includes nearly 25,000 species) 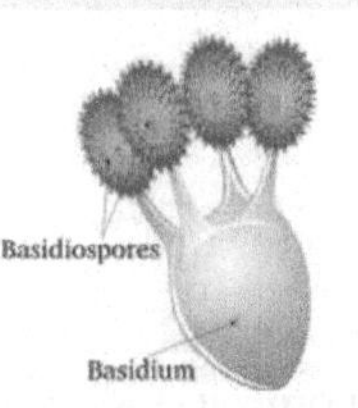	They are also called capped mushrooms and include edible mushrooms. Basidiospores formed in the basidia enable sexual reproduction.
DEUTEROMYCOTA (includes nearly 25,000 species)	This group includes other mushrooms not classified in the above groups because of their features.

SELF CHECK

KINGDOM FUNGI

A. Key Terms

Hyphae	Deuteromycota
Mycelia	Penicillium
Haploid	Antibiotic
Lichen	Zygomycota
Ascomycota	Basidiomycota

B. Review Questions

1. What characteristics distinguish fungi from other organisms?
2. How does the body of a typical yeast differ from that of a mold?
3. What is the ecological importance of saprophytic fungi? Of lichens? Of mycorrhizae?
4. Draw the life cycle of the black bread mold.
5. What measures can you suggest to prevent bread from becoming moldy?
6. Name three types of fungi that harm plants.
7. List the structural characteristics of the blue-green algae. Where are they found?
8. What is the difference between a hypha and a mycelium and between an ascus and a basidium?
9. In what ways are slime molds like true fungi? In what ways do they resemble animals?
10. What are lichens? Describe the relationship that exists in a lichen. How do they reproduce?
11. What is the importance of yeast in industry?
12. What makes fungi different from plants?
13. How do molds and yeasts differ? What are dimorphic fungi?
14. What features of fungi distinguish them from plants and from animals?
15. Discuss the major roles that fungi play in ecosystems.
16. What are the "imperfect fungi"?
17. Explain the function of hyphae.
18. How does a mushroom breathe?
19. What is the importance of yeast? And in which industries is it used today?
20. Do fungi grow everywhere in the world?
21. Which organisms have chitin in their structure like fungi?
22. What are gills for?
23. What are some products made using fungi?
24. What characteristics distinguish fungi from other organisms?
25. How does the body of a typical yeast differ from that of a mold?
26. Draw the life cycle of the black bread mold.
27. What measures can you suggest to prevent bread from becoming moldy ?

C. True or False

1. ☐ Fungi have procaryotic cells.

2. ☐ Fungi one exclusively either saprophytic or parasitic.

3. ☐ Rhizopus stalonifer is an edible mushroom.

4. ☐ Saccharomyces cerevisiae is utilized in baking bread

D. Matching

a. Lichen	( ) black bread mold
b. Ascomycota	( ) club fungi
c. Zygomycota	( ) an algae a fungi
d. Rhizopus Stolonifer	( ) saccharomyces cerevisiae
e. Yeast	( ) molds
f. Basidiomycota	( ) sac fungi

E. Fill in the blank

1. A mold consists of threads of cells called ....

2. Some hyphae are divided by walls and are ....,whereas other hyphae lack these walls and are ....

3. Yeasts reproduce asexually, mainly by ....

4. The familiar portion of a mushroom is actually a large fruiting body known as a(n).................. ....

5. The type of sexual spore produced by a mushroom is a(n)..

6. Deuteromycetes are also known as ....

F. Multiple choice

1. Fungi likes plants. Because they have...........

A) Cell membrane
B) Cell wall
C) Ribosome
D) Chloroplast
E) Root

2. Fungi are used in the production of..............

A) Food
B) Cotton
C) Cheese
D) All of them
E) None of them

3. Yeast reproduces by means of..............

A) Binary fission
B) Spore
C) Budding
D) All of them
E) None of them

4. Mushroom reproduces by means of..............

A) Binary fission
B) Spore
C) Budding
D) All of them
E) None of them

5. Fungi help in............................

A) Salt formation
B) Mineral formation
C) Oxygen formation
D) Soil formation
E) All of them

6. Which of the following is false for fungi ?

A) They are saprophytic
B) They are parasitic
C) They can participate structure of lichen
D) They are photosynthetic
E) None of them

7. Zygomycetes includes the......................

A) Bread mold
B) Sac fungi
C) Club fungi
D) Imperfect fungi
E) All of them

8. Saccharomyces cerevisae is used in the production of............................

A) Bread
B) Beer
C) Cheese
D) Soil
E) Sugar

9. Mushroom belongs to....................

A) Zygomycetes
B) Ascomycetes
C) Basidiomycetes
D) Deuteromycetes
E) All of them

10. In lichen, algae gives............. to fungi.

A) Water
B) Carbon dioxide
C) Oxygen
D) Food
E) All of them

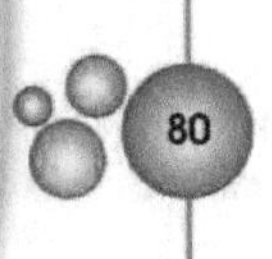

X. KINGDOM PLANTAE (PLANT

Animals and humans depend on plants for oxygen and food. Thus, plants have an important effect on the population of animals in any given place.

Plant cells contain chlorophylls (photosynthetic pigments) and carotenoids (helping pigments). Plant cells differ from those of animals due to the presence of a cellulose cell wall and the absence of centrosomes.

Furthermore, most lack true leaves, roots and stems. Alternation of generations (metagenesis) is observed in reproduction: Asexual reproduction with spores alternating with various forms of sexual reproduction according to group. In seedless plants the haploid state is longer, while the diploid state is shorter. This is the opposite of the seed plants.

Flowers serve as sexual organs and pollination is followed by development of embryo, seed and fruit. The haploid state (gametophyte) is observed only during gametogenesis. Reproduction can be either sexual or asexual.

A.

In other words, they depend on water in reproduction, while maturation takes place in various media. Mosses have more complex structures than thallophytes (algae and lichens).

(Figure 10.1). They attach themselves to the ground using rhizoids. These root-like extensions absorb water from the soil, while hair-like structures on other parts of the plant absorb water via osmosis. The vascular system is either absent or primitive.

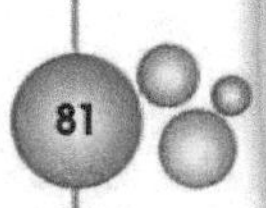

SHOW!!

Figure 10.1.: Mosses inhabit rock surfaces, moist soil and tree trunks.

ECONOMIC IMPORTANCE OF MOSSES

The high water absorption capacity of mosses, their ability to aerate the soil, and their elasticity all help to improve soil quality. Consequently they are commonly used in greenhouses and pots. In Japan they are used in parks and gardens instead of grass. They produce natural antibiotics and are not harmed by microorganisms and insects. Some species are used medicinally. The genus Sphagnum is the most widely used and has the greatest economic importance. In addition to its rich antiseptic content, Sphagnum is more absorbent than cotton, and is used as a bandage material. Furthermore, it is commonly used as an energy source in heating.

Ecologic Importance of Mosses

Mosses are one of the major components of the forest ecosystem. They play an important role in the preservation of moisture and the germination of seeds on the forest floor. Mosses can absorb 3 to 12 times their weight in water. This is important in preventing erosion.

Their female and male sex organs are archegonium and antheridium respectively. Gametophytes form gametes, whereas sporophytes form spores in sporangia. Thus there is an alternation of generations in bryophytes. Sporophytes are found on gametophytes and indeed gametophytes are the dominant form in the life cycle.

They dry out in summer and darken in color. With the return of rain, they acquire their original green color.

a. Hepaticae (Liverworts)

Gametophytes are either thallus-bearing or leafy plants.

b. Musci (Leafy Mosses)

Gametophytes are leafy plants. Sporophytes have stomata. Sporophytes display more advanced differentiation than liverworts. With the presence of an advanced photosynthetic system sporophytes can live independently.

2. [illegible]

The greatest differentiation of organs in [illegible] Found worldwide, [illegible] have vascular systems but [illegible] A vascular system and stomata are first observed in this group. Their size varies from a few millimeters to tree-like giant forms.

Ferns are photosynthetic plants displaying metagenesis as in the preceding group, but with larger sporophytes. Their vascular system is formed by tracheids. They have rhizoids (primitive roots), stems and leaves. Rhizoids absorb water and minerals from the soil. Most ferns have underground stems (rhizomes) from which roots and fronds emerge. Leaves are connected to the stem by stalks.

a. Lycopodiinae (Club Mosses)

This group has only 4 extant representative genera. Leaves lack stalks and are needle-like. True roots are first observed in this group. Some Lycopodia are used in the treatment of skin diseases, and they are used in drug production because of their drying properties.

b. Filicinae (Ferns)

Fern sporophytes are large and usually have pinnate leaves. Leaves emerge from underground stems called rhizomes. Usually spore production and photosynthesis are carried out by the same leaf. Size varies from a few millimeters to tree size. The vascular system is developed, but phloem tissue lack companion cells. Reproduction is by metagenesis.

.a. Equisetum (Horsetail)

These are found mostly in swampy and moist places. Their stems have nodes from which thin extensions spread. Green stems bear stomata. Rhizomes produce true roots.

Extracts of some species are used orally for bleeding, urinary and kidney problems.

B. SEED PLANTS

Alternation of generations is seen as in mosses and ferns, but the sporophyte is dominant in seed plants. Pollen grains are produced in large amounts on anthers. When one reaches an ovum (egg), a zygote is formed (sexual reproduction). Fertilization and development of gametophytes occur inside a flower. The gametophyte is the embryo inside the seed. Germination of the gametophyte ends the gametophytic stage, and the embryo grows to form a sporophyte plant. Two outer covers (calyx and corolla) protect the stamens (male organs) and pistils (female organs).

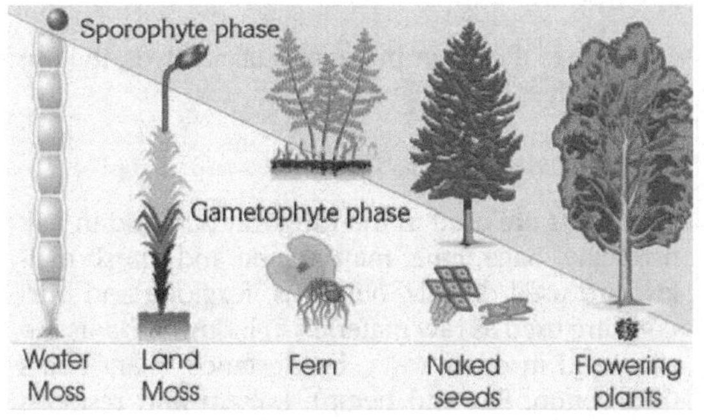

Figure 10.2.: Gametophyte and sporophyte stages in plant life cycles. Upper triangle represents the sporophytic stage and the lower represents the gametophytic stage. As seen, the gametophytic stage decreases towards the angiosperms, in which this stage is restricted to seeds.

Economical Importance of Plants

Plants are widely used in various ways.

Cellulose and paper

The most common source of cellulose is soft wood conifers. Sunflower, rice, barley, tobacco, and cotton are also important sources of cellulose, but can't be used industrially.

Oil

Many plant oils are used in food and industry. These include olive, sunflower, cotton, sesame, soy bean, peanut, pine nut, hazelnut, flax and castor bean oil.

Etheric oils

Called "essences", these chemicals are obtained from rose, thyme, jasmine, pine, lilac, lemon, lavender and violet, among others. Etheric oils are used in perfume, cosmetics, soap and drugs. They are volatile and odorous.

Dye

Due to the lower prices of artificial dyes, the use of plant dyes has declined.

Medicinal chemicals

Plants are used in the drug industry and in folk medicine. Sage, lime, mint, thyme and marsh mallow are used directly, but ferns, foxglove and burdock are used as raw materials in industry. Plants are also used in other ways. For instance, plant fibers (like cotton, flax and hemp), balsam and resinous plants (gum tree, pine) are used in various industries.

If a flower has both male and female organs, then it is called a hermaphrodite. If the male and female organs are found in the same plant but in different flowers, the plant is called monoecious. Pines are an example of monoic plants. In some plant species, like willow and mullberry trees, male and female flowers are found on different plants, and these are called dioecious plants.

In the higher plants, dependence on water decreases. As a result, the gametophytic stage is decreased (*Figure 10.2*).

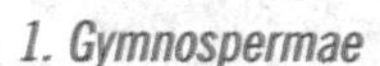

1. Gymnospermae

Mostly closely branched trees and rarely bushes, gymnosperms have seeds that are not closed inside carpels, hence, the name gymnosperms.

Gymnosperms are woody plants with secondary growth. Their xylem lacks tracheids. Scale or needle-shaped leaves are renewed constantly and thus gymnosperms are evergreen.

Gymnosperm flowers are single sexed; either male or female. Pollen is transferred by the wind, or rarely by insects. Flowers are cones. Seeds spread when the cones open. Germination of seeds produces new trees.

Figure 10.3.:Pine cone. Cones are the sexual organs of gymnosperms

The leaves' shape, outer protective layer, inner stomata, and resin channels all demonstrate that gymnosperms are adapted to dry conditions.

There are 63 genera and 722 species of gymnosperms.

Coniferophyta (Conifers)

Conifers' leaves are needle shaped. Most of them, such as pine, cedar, spruce and fir, are monoic plants. They are very important ecologically as they provide nutrition and shelter for many animals, and their roots prevent erosion. They are used in the production of wood, paper, resins, naphtha oil, etc.

a. Fir (Abies)

Firs are evergreen plants reaching as much as 80 m. in height. Their cones are more vertical than those of other plants. Examples of firs include Taurus fir, Mt. Ida fir and Black Sea fir.

Abies spp.

b. Spruce (Picea)

Spruce are found in high mountains. They require high moisture. Leaves are quadrangle in cross section and emerge one by one. The tips of hanging cones are pointed. Shallow roots are weak, especially vulnerable to strong winds.

Picea spp.

c. Cedar (Cedrus)

Cedar leaves are found in clusters and the minute cones are vertical. They are found from the Mediterranean Sea to India. Examples are Atlas and Lebanon cedars.

Cedrus libani

d. Pine (Pinus)

Pines are the most common conifers. Various species form large forests in the northern hemisphere. Examples include black, red, and yellow pines.

Pinus pinea

e. Cypress (Cupressus)

Cypress are brush-shaped trees reaching 20-30 m. in height. Their distinct cones are round. They are found in higher mountains and cool areas. Needle-shaped leaves become scaly when mature.

Cupressus spp.

f. Juniper (Juniperus)

Junipers are bushes and trees. They are wide-spread in the Mediterranean region and in the Taurus Mountains.

Juniperus spp.

There are many phyla in this group. These include cycodophyta (100 species, mostly extinct), Gingkaphyta (1 living species), and Gnetophyta (which resembles angiosperms).

2. Angiospermae (Flowering plants)

This is the most diversed group with about 12,500 genera and 265,000 species. The majority of cultivated plants belongs to this group.

Somes differences from gymnosperms:

- Seeds are contained inside ovaria formed by carpelles. Ovaria form fruit.
- Pollen can not reach the ovary directly. Upon reaching the stigma, the pollen forms a pollen tube. The pollen reaches the ovum (egg) through the pollen tube and fertilizes the egg.
- Gametophytic stage is shorter.
- Double fertilization occurs. One sperm fertilizes the egg while another fuses with the polar nuclei to form the endosperm.
- May be herbaceous plants as well as woody plants.
- Vascular system has tracheids and companion cells.
- Flowers have different colors and fragrances to attract insects.

Angiosperms are grouped into two classes (Monocotyledanae and Dicotyledanae) according to the number of cotyledons.

The main differences between the classes (Figure 10.4).

- Venation of leaves: Dicot leaves have a network of veins, while parallel venation is observed in monocot plants.
- Flowers: Organs (petals, sepals, stamens) are found in series of 4 or 5 in dicots. Monocot flower organs are usually found in series of 3.
- Root: Dicots have one primary root and thinner secondary roots developing from this root. Monocots have adventitious roots and all roots are the same size.

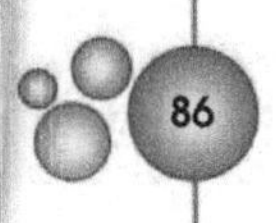

- Stem: The dicot vascular system is often arranged in a circle, whereas vein distribution in monocots is irregular. Furthermore, dicots have cambium tissue in veins, monocots do not.
- Cotyledon: Dicots have two cotyledons, but monocots have only one.

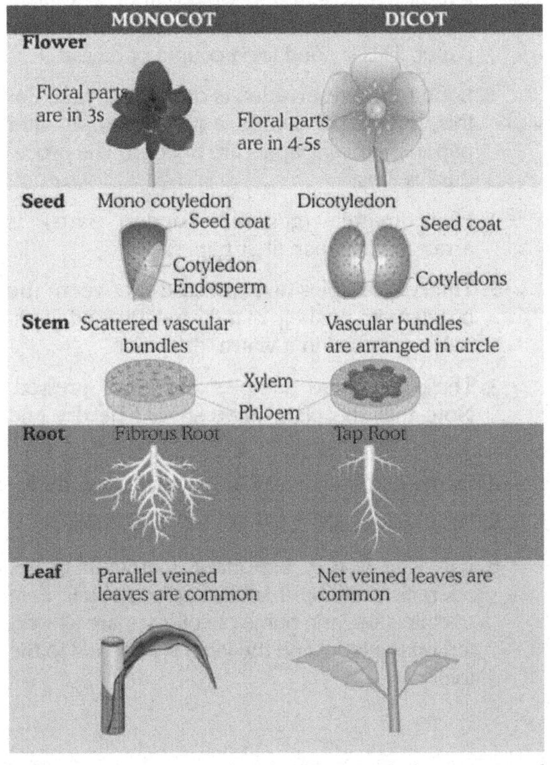

Figure 10.4.: Comparison of flower, seed, root and leaf in monocots and dicots.

2.1 Monocotyledon

As mentioned before, these are grassy plants, e.g. wheat, corn, palm tree, etc.

a.Couch grass (Agropyrum repens)

Being a Gramineae(grass family) member, this plant has rhizomes. These are collected in spring and autumn for medicinal purposes.

Agropyrum repens

b. Garlic (Allium sativum)

A member of the lily family, garlic is anti-hypertensive and appetizing. Volatile oils are antibacterial and applied to the skin for many diseases, like scabies and alopecia.

c. Onion (Allium cepe)

Onion is another therapeutic member of the lily family.

d. Purple Iris (Iris germanica)

A member of the family Iridaceae, iris is used for therapy of injuries.

Iris germanica

2.2. Dicotyledons

Most of the higher plants are dicots.

a. Walnut (Juglans regia)

Walnut is used for the treatment of constipation and rheumatism. It is also used as a dye, the fruit is eaten and the timber is valuable.

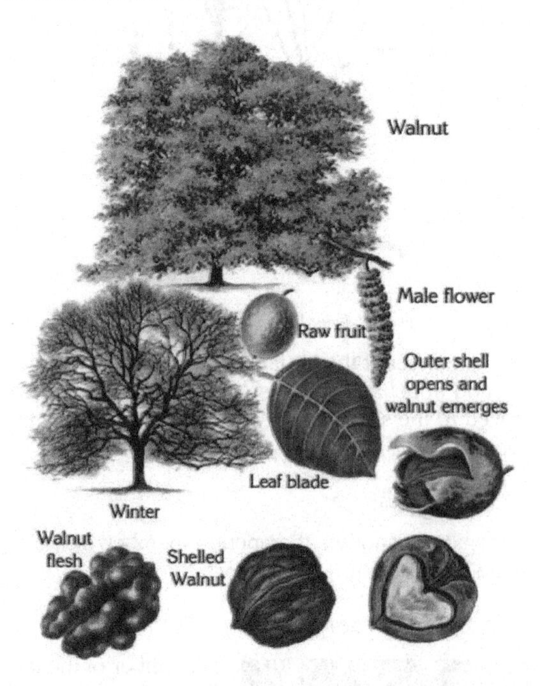

b. Lemon (Citrus medica limonia)

The Mediterranean lemon is in the citrus family and is rich in vitamin C.

Preservation of plant samples

Plant matter can be preserved fresh or dry.

- For short term preservation, use a sticky, transparent plastic film and white paper. The plant matter is positioned between the film and the paper. This is good for a couple of days.
- Long term preservation is done by drying. For this, 2 plywood boards, a piece of rope, filter paper and newspapers are needed. The procedure is thus:

1. Plant matter (generally upper parts) is arranged between filter papers.
2. Thicker samples are pressed between the boards with the help of the rope. Pressed materials are stored in a warm, dry place.
3. These prepared samples are then pressed. Note that the plant press should be dry and hot.
4. Damp newspaper is changed every few days.
5. Dried plants are inserted in a paper folder.
6. Labelling is very important. Common name, scientific name and family name, habitat, date of collection, and name of collector are all written on the label and the label is attached to the folder.

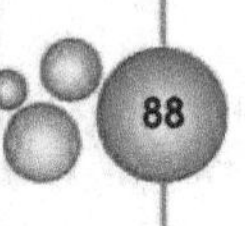

c. White willow (Salix alba)

The boiled extract of willow bark is appetizing and is used medicinally, e.g. for the treatment of constipation and fever.

d. Linden (Tilia spp.)

Flower tops are used as a tea in folk medicine.

e. Mulberry (Morus nigra)

Syrup from the fruit is used to treat throat diseases, while the bark of the roots and stems is anthelmintic.

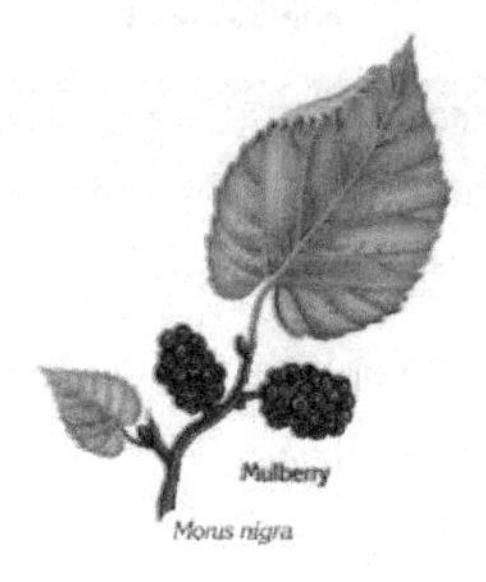

f. Olive (Olea europea)

Besides being eaten, the fruits are sources of olive oil. Soap and salve are made from the fruit. Bark and leaves regulate hypertension and have anthelmintic effects.

g. Tea (Thea cinensis)

Tea originated in Ceylon (Sri Lanka). It is widely cultivated and is popular as a hot or cold drink.

h. Tobacco (Nicotiana tabacum)

Tobacco leaf extracts are used against garden insect pests. The leaves are smoked, and cooking oil is obtained from the seeds.

ı. Some other dicots

Raspberry, pea, chickpea, mint, rose, cherry, peach, almond, parsley, etc. are all dicots commonly used by humans.

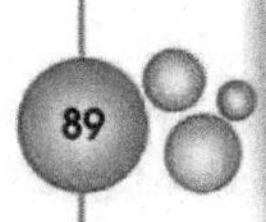

SELF CHECK

PLANTS

A. Key Terms

Bryophyta
Hepoticae
Musci
Pteridophyta
Lycopodinal
Filicinae
Equisetum
Seed plants
Fern
Gymnosperm
Angiosperm
Coniferophyta
Polen
Cotyledone
Seedless plants

B. Review Questions

1. List the adaptations of Bryophytes that are lacking in algae. What adaptations do ferns have that both algae and bryophytes lack?
2. How does heterospory modify the life cycle?
3. What kind of plants lack roots, stems, and leaves?
4. What is the function of vascular tissue?
5. What are the differences between mosses and ferns?
6. What makes ferns different from vascular plants?
7. What is double fertilization?
8. How are flowering plants different from gymnosperms?
9. How does the gymnosperm life cycle differ from that of flowering plants?
10. What are the two classes of flowering plants, and how can one distinguish between them?
11. How are cones and flowers alike? How are they different? (Hint: You should consider microspores, megaspores and seeds).
12. Why is a moss plant restricted to a height of less than about 15 cm?
13. Why must mosses and liverworts (phylum Bryophyta) always live in close association with water?
14. What are spores? How are they formed in mosses and in ferns?
15. In what ways do ferns resemble seed plants? In what ways are they different?
16. Differentiate between the monocots and dicots.
17. What is alternation of generations in plants?
18. Why do mosses need water to reproduce?
19. Which generation is dominant in the life cycle of ferns?
20. Outline the classification of land plants.
21. What are the main properties of mosses?
22. What are the main properties of liverworts ?
23. What are the main properties of hornworts?
24. Do the structures of liverworts and hornworts have any relation to their names?
25. Write down the main properties of mosses.
26. What does bryophyta mean?
27. What is the difference between bryophytes and tracheophytes?
28. What are the similarities between mosses and algae?
29. Why are algae not classified together with mosses in the same kingdom?
30. Why can't the bryophytes grow as high as ferns?
31. What is thallus? Which bryophyte has it?
32. What are the similarities and differences between rhizoids and roots?
33. Draw a phylogenetic tree that shows the relationship of the various plants.

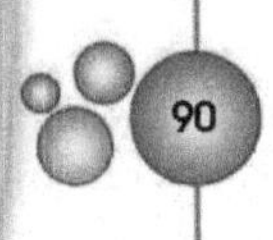

C. True or False

1. ☐ Grassy plants are monocotyledone.
2. ☐ Dicots have only one cotyledone.
3. ☐ Gametophytes reproduce by means of their gametes.
4. ☐ Amphibians at plant world is bryophyta.

D. Matching

a. Bryophyta	( ) Alternation of generation
b. Metagenesis	( ) Root-like extensions
c. Rhizoids	( ) Land mosses
d. Monocot	( ) Single cotyledone
e. Dicot	( ) Double cotyledone
f. Stamen	( ) Male organat flowering plants.
g. Pistil	( ) Female organ at flowering plants

E. Fill in the blank

1. The ...., which include mosses, liverworts, and hornworts, are small plants that lack a vascular system.
2. The waxy layer that covers aerial parts of plants is the ....
3. The openings in the plants that allow gas exchange for photosynthesis are called ....
4. The leafy green moss plant is the .... generation.
5. The flattened, leaflike body form of many liverworts is called a(n)..................................
6. .... are bryophytes with chloroplasts that are similar to those of certain algae.

F. Multiple choice

1. Which of the following plant group carry its embryo on its leaves ?

A) Bryophyta B) Ferns C) Monocotyledone D) Dicotyledone E) All

2. Which of the following organism is involved in the production of agar ?

A) Phaeophyta B) Rhodophyta C) Chlorophyta D) Gymnospermae E) Angiospermae

3. I. Monera
II. Protista
III. Plants

Chloroplast are found in which of the above kingdoms?

A) I only B) II only C) III only D) I and III E) II and III

4. Which of the following is incorrect concerning green algae and higher plants?

A) Both are usually multicellular and motile.
B) Both capture energy by utilizing the same pigment.
C) The cell walls of both are mostly composed of cellulose.
D) Both usually store glucose as starch.
E) Both of them are photosynthetic.

5. An algae is marine, multicellular, lives in fairly deep water, and it has phycoerytherin. It probably belongs to which group?

A) Rhodophyta
B) Phaeophyta
C) Chlorophyta
D) Dinoflagellata
E) Chrysophyta

XI. KINGDOM ANIMALIA (ANIMALS)

Animals are some of the most common organisms found from the oceans to the high mountains. Except for some animals like sponges, most animals are easily differentiated from other groups (e.g. plants or fungi). For definite separation, the characteristics used are these:

- All animals are eukaryotes and metozoan (multicellular).
- All animals are heterotrophs. Most digest their food internally.
- Most animals are motile during at least a certain period of their life.
- Many animals have a well-developed nervous system and sensory organs.
- Most animals reproduce sexually. The organism develops from a zygote through embryonic stages.
- Animals possess organs for respiration, excretion, and circulation, except for primitive phyla which use diffusion for these functions.

Over one million species are already known to science, and millions more are expected to be discovered in the future. A minority (5%) of all animals are vertebrates. Vertebrates are animals that have a backbone. Those without a backbone are called invertebrates. Examples of the latter, much larger group are sponges, jellyfish, insects, snails, etc.

A. CLASSIFICATION OF ANIMALS

a. Cellular Organization

All animals are made up of cells. Organized groups of cells form tissues, organs and systems. The diversity and organization level of cells become more complex in higher groups. Different functions, e.g. respiration, reproduction, feeding and protection, are performed by different cell groups. Organ systems form the organism and work in harmony. Diversity and organization of cells is a factor in the classification of animals.

b. Symmetry

Except for a few groups, the majority of animals exhibit bilateral symmetry and can be divided into parts e.g. head, torso and limbs. For identifying various fea-

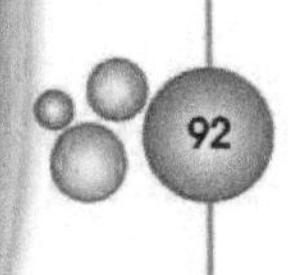

tures of such animals, the terms anterior, posterior, dorsal and ventral are used. In the same way, for understanding the inner structures, vertical, transversal, and horizontal planes (sections) are defined (Figure 11.1).

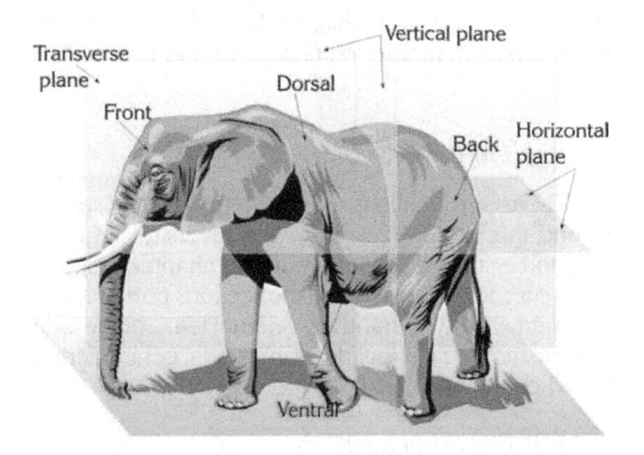

Figure 11.1.: Terms used for determination of symmetry in animals

Symmetry is the presence of one or more planes that divide an organism into identical sections. Many unicellular organisms are asymmetrical (no equal places). In metazoans, radial (cnidaria) and bilateral (many groups) symmetry is seen (Figure 11.2).

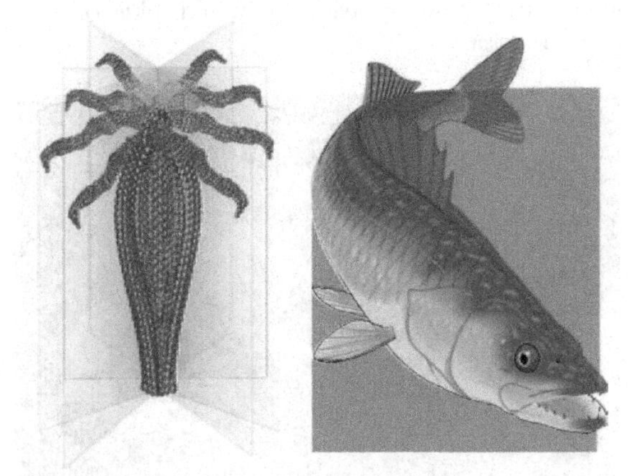

Figure 11.2.: (Left) Animals exhibiting radial symmetry can be divided by any number of planes passing though the body axis. (Right) In bilateral symmetry, an animal can be divided by only one longitudinal plane passing through the axis.

Organisms with radial symmetry can be divided into equal pieces by any number of planes passing through the main body axis. Bilaterally symmetric animals can be divided into two pieces by a single longitudinal medial plane.

Generally, slow or sessile animals are radially symmetric, whereas faster animals have bilateral symmetry. Examples of the former are jellyfish, hydra and some sponges. All higher organisms including humans are bilaterally symmetric.

c. Embryonic Layers

At the beginning of embryonic development, cells are arranged in two or three layers from which organs and systems will be produced. Two layers (ecto- and endoderm) are seen in sponges and cnidarians. All higher animals have an additional intermediate layer (mesoderm).

d. Body Cavity

In many metazoan groups there is a single body cavity with a combined mouth/anus (like hydra). Higher animals have two body cavites; a true coelom, and a digestive cavity with separate oral and anal openings.

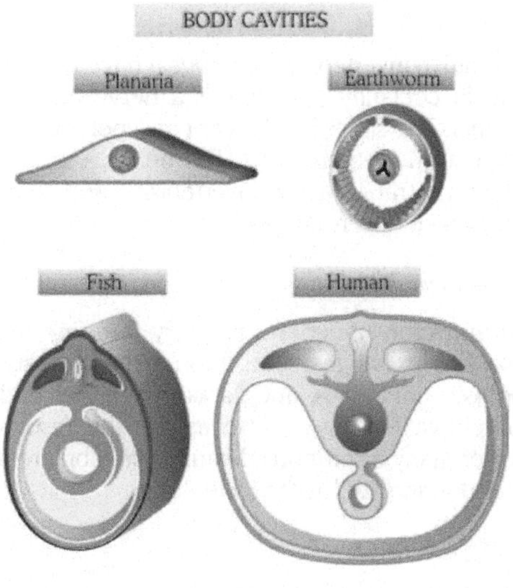

BODY CAVITIES

Planaria

Earthworm

Fish

Human

Figure 11.3.: In the classification of animals, body cavites are one of the most basic criteria. In earthworms and simpler organisms, the only body cavity is the digestive tract. In higher groups, including mollusks, there is an additional cavity called the coelom.

e. Segmentation

Some animals are segmented (annelids, etc.) while some lack segments (vertebrates). Segments may be identical, although different in size.

f. Skeleton

Many animals have hard tissues functioning as support structures. These structures differ in their chemical composition, e.g. bone, cartilage, or structures made of silica, calcium carbonate, keratin or chitin. Chemical composition of the skeleton is important in classification. Bony or cartilaginous skeletons are found in vertebrates, whereas invertebrates have other sketetal forms. Intermediate animals between invertebrates and vertebrates are called primitive chordates and their vertebrae (backbone) are called notochords.

g. Circulatory System

The circulatory system of animals may be one of two types; open or closed. In a closed circulatory system, blood flows inside vessels. Hemoglobin carries O_2 and CO_2 in the blood and is found in the plasma (in annelids) or inside erythrocytes (in all vertebrates).

h. Nervous System

Primitive invertebrates (sponges) have no nervous system. In coelanterates there is a nerve net (diffuse ganglions). The simplest, "ladder-type" nervous system is first seen in the flatworms. The nervous system lies along the ventral surface in invertebrates and along the dorsal surface in vertebrates.

i. Respiration

Respiration mechanisms vary among animal groups. Gas exchange occurs through the body surface in sponges, cnidarians, and flatworms. Fishes and larval amphibia have gills. Trachea are the respiratory organs of many arthropods. Mature amphibia, reptiles, birds and mammals breathe with lungs.

j. Excretory System

According to the level of organization, animals have different excretory systems. Poriferans and Cnidarians have no excretory system and their wastes pass by diffusion. Protonephridia (flatworms), nephridia (annelids), and malpighian tubules (arthropods) are common excretory organs. Kidneys are the excretory organs of vertebrates, but they differ in various vertebrate groups. A mesonephros system is present in the embryos of all birds, reptiles, mammals, mature fishes and amphibia, while a metanephros type is found only in mature reptiles, birds and mammals.

B. ANIMAL KINGDOM

There are millions of different animal species around the world. Because of this diversity all groups have to be studied separately. In this chapter, kingdom Animalia will be divided into the sections porifera, coelenterata, platyhelminthes, nemertea, nematoda, rotifera, mollusca, annelida, arthropoda, echinodermata and chordata.

1. Porifera (Sponges)

Porifera means "having pores", which describes their perforated, sac-like bodies. These are mainly marine animals with the exception of a few freshwater groups. There are about 10,000 species alive today. They are sessile and mostly attach themselves to hard surfaces like rocks and shells. They may be flat, ball- or vase-shaped. Size varies from a millimeter to 200 cm in length. They may be yellow, grey, red, blue or even black in color.

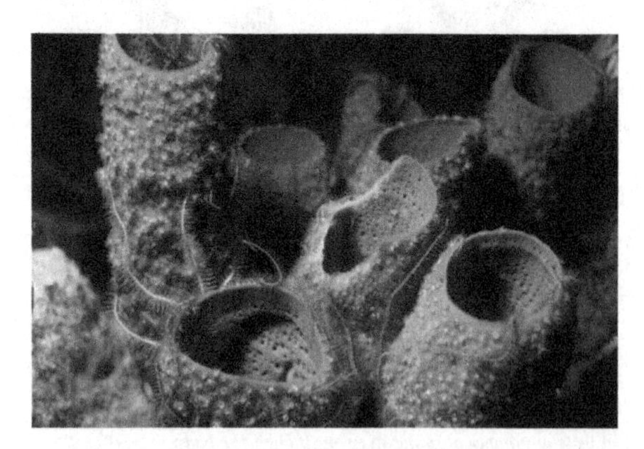

Cell differentiation in this group is incomplete. There is no tissue organization or sensory, nervous or muscular system. As a result, they can't move. They are

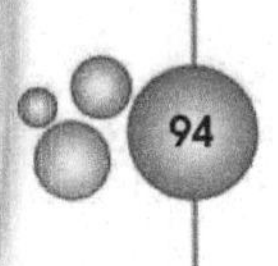

bilayered, radially symmetric or asymmetric. There is a central cavity (in sponges) opening to pores (ostia), and one or more mouth openings covered internally with flagellated collar cells called choanocytes. The choanocytes circulate the water inside the cavity and trap food particles or plankton. Choanocytes are also responsible for respiration and excretion.

The outer layer is composed of epidermal cells and porocytes, or ostia (sing. ostium), and is covered with spicules. Between the two layers is a gelatinous intermediate layer called the mesohyl. The mesohyl includes amoebocytes which transport food or function in digestion. Sponges differ on the basis of the type of skeleton they secrete. Some species secrete a skeleton of calcium carbonate; some have a skeleton made of spicules containing silicon, and some others have skeletons of a protein known as spongin. These spicules support the mesohyl and usually extend out from the epidermis and sometimes form a crown around the osculum (Figure 11.4).

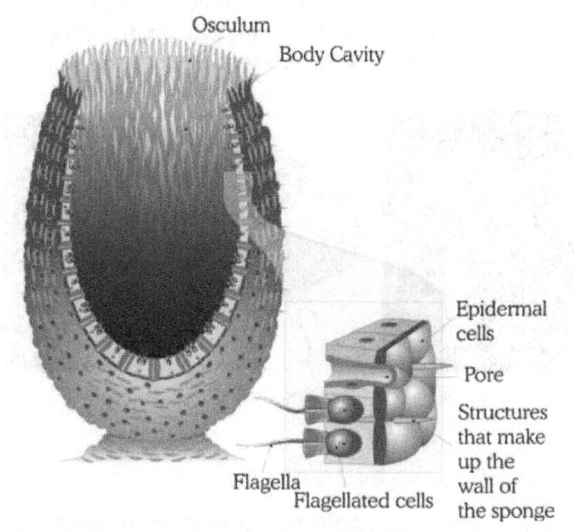

Figure 11.4.:Structure of sponges. Pores connect the inner cavity to the outer surface. A section from the body wall shows spicules, choanocytes and epidermal cells with pores.

In Porifera, reproduction is sexual or asexual. Budding or regeneration are asexual forms of reproduction. In the latter, broken fragments form a new sponge on or near the parent sponge. Most sponges are hermaphroditic: male and female gametes are produced by the same sponge. But some others may be monoic, that is the gametes are produced by different individuals.

2. Cnidaria (Coelenterates)

Coelenterate means "hollow (animal)" in Greek. There are over 10,000 species in this group. Most are marine but there are a few freshwater representatives. They live solitarily or in colonies.

In general they have two morphs; polyp and medusa (Figure 11.5).

- Polyp: These are sessile forms, and thus fixed to the ground. Their bodies are tubular with an oral aperture functioning as mouth and anus. Around the "mouth" there are many stinging tentacles.
- Medusa: These swim freely in the water and have a flat, umbrella like body. Their body is jelly-like. Except for their shape and some other features, their general anatomy resembles that of the polyps.

Some groups have both forms (medusa and polyp) in their lives. But some lack one stage, and are either polyps or medusae their entire life.

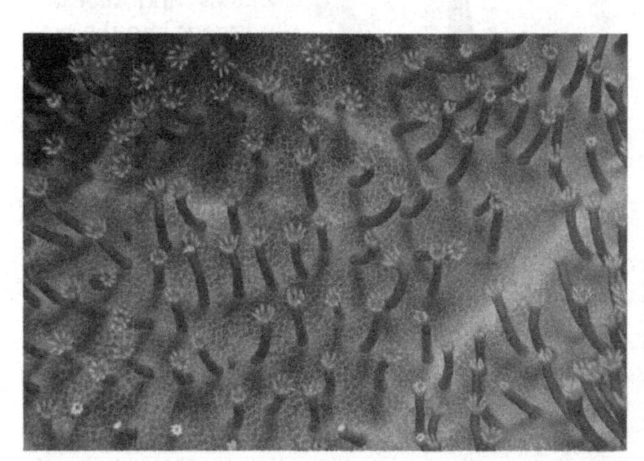

Figure 11.5.:Polyp and medusa types in cnidarians. Polyps live attached to the bottom (benthic), while medusae live freely (pelagic). Medusae bud off from the polyps. Then medusae produce polyps by sexual reproduction. Thus there is metagenesis in Cnidaria.

There are 3 classes in the phylum Cnidaria (Coelenterata). These are Hydrozoa, Scyphozoa and Anthozoa.

a. Hydrozoa (Hydra)

The hydra is the typical representative of this group. Hydra are small polyps (appr. 5-6 mm in length) that live in streams, ponds and lakes. Aboral side is attached to the substrate. The mouth is positioned on the upper tip. There are 6-8 tentacles around the mouth active in movement and feeding.

There are two dermal layers in hydra: ectoderm and endoderm. Ectoderm cells are cubic and endodermal cells are cylindrical. There is an intermediate gelatinous layer, mesoglea, which is the common product of the two layers.

Some ectoderm cells are differentiated into muscle, neural, sensory and secretory cells. Cnidocytes, which contain stinging nematocysts and function in feeding and protection, are another type of cell.

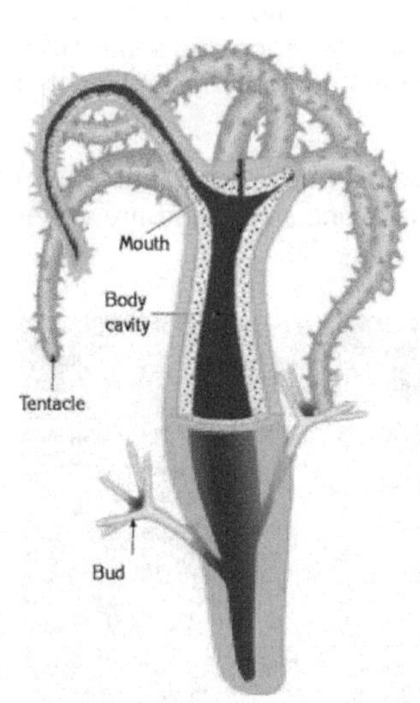

Endodermal cells, which form the gastrodermis, coat the gastrointestinal cavity, or enteron. Some endodermal cells secrete mucus and digestive enzymes into the cavity. Flagellated cells lining the body cavity circulate water inside the cavity and function in digestion and absorption of food. As in the ectoderm, there are sensory and nerve cells in the endoderm. Digestion begins in the cavity and is completed inside the cells. But there is no respiratory, circulatory or excretory system in hydrae. These functions are carried out via diffusion. However, the first nerve cells in the animal kingdom are found in this group. These cells form a basic and diffuse system called a nerve net.

Although sessile they are able to move short distances. They extend their tentacles to plant stems or stones, and either push themselves to another place or move with a series of somersaults.

They reproduce asexually by budding or regeneration, but they can also reproduce sexually. Sexual organs can be seen as small buds on the tube-shaped body. They may be hermaphroditic or monoic.

Colonial forms constitute an important part of this group. There is a remarkable organization in these animals. Some individuals of the colony are specialized for feeding, while some others lack tentacles and even a mouth, serving only in reproduction.

b. Scyphozoa

All are umbrella-shaped marine organisms. Their polyp forms are usually reduced. The jellyfish (Aurelia aurita) is an example of this group. They may attain large sizes (in the genus Cyanea, up to 2 m in diameter). Coloration is translucent bluish or even pinkish. They are jell-like: 96% water by weight. They are mostly pelagic. They live near the water surface. They have simple eyes (rhopalia) and organs for balance and chemical reception. In Aurelia, the umbrella has 8 lobes. Lobes carry peripheral tentacles. In the center, there is a mouth and reproductive organs arranged around it. Scyphozoans have separate sexes.

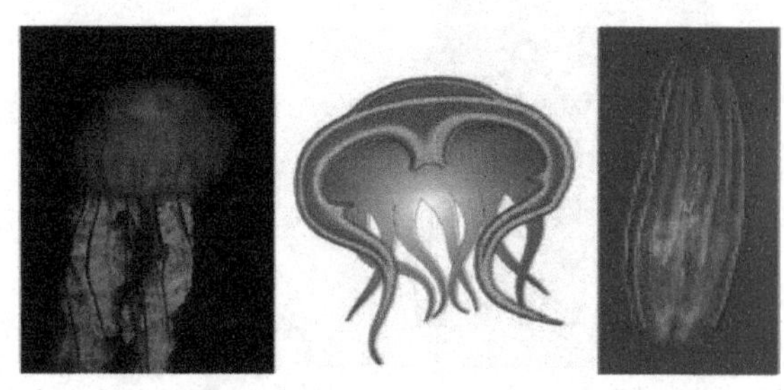

Figure 11.6.: Photographs and schematic view of jelly fish.

c. Anthozoa (Sea Anemones and Corals)

Sea anemones and the colonial coral groups are anthozoans. This group is remarkable for lacking a medusa generation. Radial symmetry is seen in some members (sea anemones). Most are colonial and sessile. Tentacles encircling the mouth are present in general and they capture small plankton, and even fish and sea invertebrates. They usually reproduce sexually. A planula larva, a ciliated form of larva seen in many cnidarians, attaches itself to a hard surface, and a colony develops by budding (in colonial species).

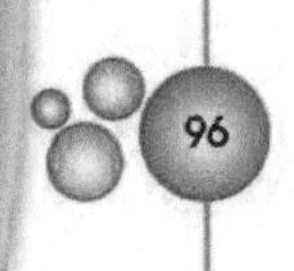

Figure 11.7.: A photograph and schematic view of a sea anemone.

Calcium carbonate in the skeletal structure provides a hard covering in coral colonies. The accumulation of the skeletons of dead colonies creates coral reefs and atolls especially in the Pacific Ocean. Red coral is used in jewelry.

Zooxanthellae are photosynhetic symbionts of corals and play an important role in the development of coral reefs. Zoochlorallae are similarly found in Hydrozoans and sometimes give them a green colour.

3. Ctenophora (Comb jellies)

These are all very delicate and often luminescent marine organisms. They acquire their names because of their apperarence: They resemble cnidarian medusa covered with comb like cilia. There about 100 species.

There is a thick mesoglea between the endoderm and ectoderm. Some have a pair of tentacles. Their sensory organs provide balance.

4. Platyhelminthes (Flat-worms, Helminths)

These are flattened, soft-bodied organisms, and are the first animals with bilateral symmetry. There are 20,000 known species. They are mostly aquatic (marine or freshwater), but there are some terrestrial species of moist soil, and many parasitic species are found in various organisms. Parasitic species often lack digestive and sensory organs. They are mainly hermaphroditic.

There are 3 main classes:

1. Turbellaria-Turbellarians
2. Trematoda-Flukes
3. Cestoda-Tapeworms

a. Turbellaria (Turbellarians, Planarians)

One of the best represented groups, Planaria are found under stones in streams (Figure 11.8). They feed on plant material and small animals. They are so flat that there is no need for a respiratory system, and gas exchange is by diffusion. Their size varies between 0.5-3 cm. Their outer surface is covered with ciliated epithelium. They move by means of cilia. There are two eyes on the head. There is a single, ventral opening in the center of the body functioning as mouth and anus. There are two nerve cords in a ladder system. Protonephridia (urinary organ) with flame cells constitute the excretory system. They may reproduce sexually or asexually. The ability of planaria to regenerate is relatively high. Every cut part of a planarian is able to turn into a complete animal.

Planaria are hermaphroditic. In sexual reproduction, cross-fertilization occurs during copulation.

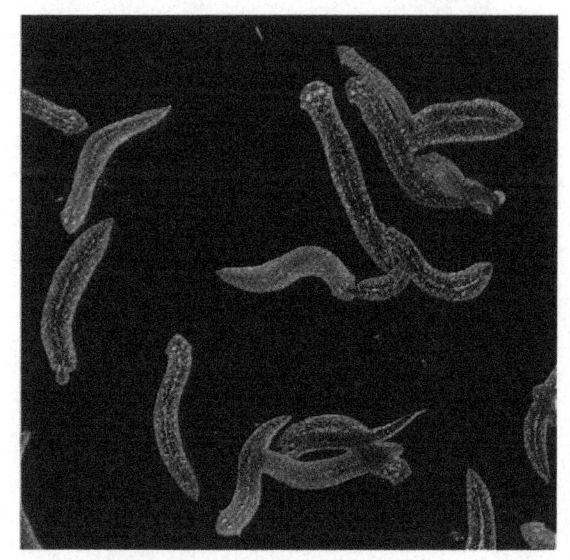

Figure 11.8.: Planarians are a flat-bodied, freshwater species.

b. Trematoda (Flukes)

Flukes are found as parasites of vertebrates and humans. They have a branched gastrovascular cavity, and oral and ventral suckers. In humans they live in hepatic veins and bile ducts. Blockage of bile ducts results in jaundice. The parasites destroy the liver, but

WORMS

Cestodes live as parasites in the bodies of probably every kind of vertebrate including humans. Their bodies consist of two parts: the scolex (a head-like structure) and proglottids (body segments). On the tip of the scolex there are often suckers and sometimes hooks. This part connects the parasite to the host. The first proglottid segments are smaller, but the posterior ones become larger. Medium-aged proglottids have fully-developed genital organs, but when they become older their uteruses are filled with fertilized eggs. Eggs are released from old proglottids shed from the worm. Development in flatworms is completed in two stages: cysticercus and adult stages. These stages live in different host organisms and are named according to the host.

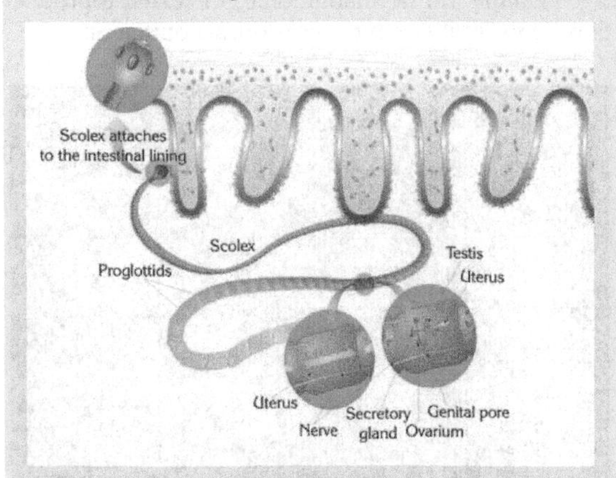

Taenia saginata (Cattle tapeworm)

When the egg-filled proglottid is passed (with feces), eggs are transfered to the environment. While eating grass, cows also ingest the eggs which then hatch in the stomach. Through the blood, the larvae invade muscles where the cysticercus larva develops and forms a cyst. If someone eats undercooked meat containing a cyst, the cyst opens in the stomach. Cysticercus adheres itself to the gut surface. The maturation period is about 2.5 months.

Taenia solium (Pig tapeworm)

The scolex has 4 suckers and a rostellum. The worm is about 3-3.5 m long. The intermediate host is a pig. The lifecycle is similar to that of the beef tapeworm. Maturation period of cysticercus inside the human intestine is about 3 months.

Taenia echinococcus (Dog tapeworm)

The adult stage is in a wolf, jackal, cat or dog (definitive hosts), cysticerus stage is in a pig, cow, sheep or human (intermediate hosts). The cyst grows in a few years, and after 15 years may be as large as a child's head. This requires surgical removal, but late diagnosis leads to death. Removal of the cyst without bursting it is very important.

Diphyllobothrium latum (Fish tapeworm)

Adults are found in the intestines of cats, dogs, pigs and humans. It is the largest of human parasites (exceeding 15 m.). The cysticercus stage passes inside fish.

Every tapeworm lays over 200,000 eggs. Eggs are ingested with unwashed food, hatch, and larvae pass into the intestine. The larvae then pass to the circulatory system, lungs and return to the intestines where they mature.

Ascaris lumbricoides (Intestinal roundworm)

There is no intermediate host. The adult form is found in the intestines of humans and pigs. Male is 15-25 cm, female is 20-40 cm long. Dirty hands and food are the means of being infected with the eggs.

Ancylostoma duodenale (Hookworm)

A blood-sucking parasite of the human intestine, hookworms cause hemorrhage and finally anemia. Length is about 1 cm. Characteric hooks on the head are the reason for the name.

Oxyuris vermicularis (Roundworm)

It prevents development in children and is found in the small intestine (in youth) or large intestine (in old people). Females leave the anus at night and lay eggs on the skin. The eggs cause an itchy feeling in children. After scratching the itchy area, autoinfection occurs unless the hands are washed.

Trichinella spiralis (Trichina worms)

Cysts develop in pork. The larvae penetrate the intestinal wall and enter the circulatory system. Through the blood, larvae reach skeletal muscles where they settle (muscle trichina). This causes severe pain. Completion of development happens inside another host. There they become intestinal trichina larvae and mate.

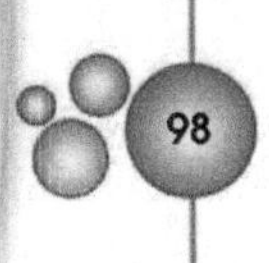

COMPARISON OF SOME INVERTEBRATES					
PHYLUM	**PORIFERA**	**CNIDARIA**	**PLATYHELMINTHES**	**NEMERTEA**	**NEMATODA**
Sample organisms	**Sponges**	**Hydra, Jellyfish, Coral, Sea Anemones**	**Planaria, Tapeworm, Flukes**	**Ribbonworms**	**Roundworms**
Organization level	**Many loosely connected cells**	**Tissue**	**Organ**	**System**	**System**
Symmetry	**Radial or asymmetric**	**Radial**	**Bilateral**	**Bilateral**	**Bilateral**
Digestion	**Intracellular digestion**	**Gastrovascular cavity with one opening, intracellular and extracellular digestion**	**Gastrovascular cavity with one opening**	**Digestive cavity including two openings: Mouth and Anus**	**Digestive cavity including two openings: Mouth and Anus**
Circulation	**Diffusion**	**Diffusion**	**Diffusion**	**Two parallel vessels, no heart; hemoglobin in blood cells**	**Diffusion**
Gas Exchange	**Diffusion**	**Diffusion**	**Diffusion**	**Diffusion**	**Diffusion**
Excretion	**Diffusion**	**Diffusion**	**Protonephridia and flame cells**	**Two horizontal excretory canals and flame cells**	**Excretory vessels**
Nervous system	**Cytoplasmic stimulation**	**A simple nerve net**	**A simple brain, ladder nerve system and simple sense organs**	**A simple brain, two nerve cords and simple sense organs**	**A simple brain, a dorsal and a ventral nerve cord and simple sense organs**
Reproduction	**Budding; Sexual and Asexual (Most hermaphorodite)**	**Budding, Sexual and Asexual (separate sexes)**	**Asexual by dividing into two; Sexual (hermaphorodite)**	**Sexual or Asexual with Segmentation (separate sexes)**	**Sexual (separate sexes)**
Support and Movement	**Support by spicules which are needles of calcium carbonate; contractile cells provide movement**	**Mesoglia, Calcium skeleton (corals) and gastrovascular cavity fluid provides support (Hydrostatic skeleton); Conractile cells**	**Developed Muscular tissue and other tissues provide support**	**Tissues provide support; Muscles and Cilia provide movement**	**Thick Cuticula, Pseudocoelomic fluid (hydrostatic skeleton); Body muscles**
Environ ment and life style	**Aquatic (mostly marine); ciliated; motile larva but sessile adult**	**Aquatic (mostly marine); Polyp and Medusa; some in colonies; Feed with cniodocyst and tentacles**	**Aquatic conditions, some in humid places; mostly carnivores, some parasites**	**Mostly marine; Mostly carnivorous; use proboscis for feeding and protection**	**Land, Sea and freshwater; Carnivores, Scavanger, and Parasites**

therapy is possible. They are either hermaphroditic or have separate sexes. Examples of the group are the great liver fluke (Fasciola hepatica) and the lancet liver fluke (Dicrocoelium dendriticum).

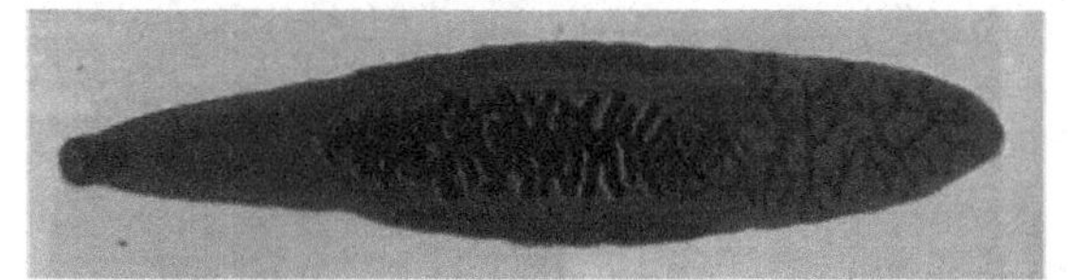

Figure 11.9.: Liver flukes are parasites of vertebrates and humans

c. Cestoda (Tapeworms)

About 1000 species are exclusively endoparasites of vertebrates or humans. Their body is flat, slender and ribbon-like. These are all hermaphroditic animals and reproduce with cross-fertilization. Examples are beef tapeworm, pork tapeworm, dog tapeworm and fish tapeworm.

5. Nemertea (Nemertins)

This small phylum consists of 900 species. Except for a few species living in moist soil or freshwater, they are completely marine. They are usually round, cylindrical, flat or ribbon-like. Size ranges from 5 to 200 cm. They are free-living animals. Coloration may be bright, even red or green, with black spots or lines. A characteristic organ, the proboscis, secretes mucus which helps in prey capture. They are the first in the animal kingdom to have separate anus and mouth. Also there is a simple, closed circulatory system, but no heart. Blood is pumped by the muscles.

6. Nematoda (Roundworms)

Nematodes are very important ecologically. Mostly they inhabit sediment layers in water (both marine and freshwater), and are abundant in soil. Their importance comes from their major role as decomposers. Their bodies are long, thin and pointed at both tips. Sense organs are not developed and they have separate sexes. There are about 12,000 species.

Although there are many free-living species, some species are plant or animal parasites. Parasitic roundworms obtain nutrients and oxygen from their host.

COMPARISON OF SOME INVERTEBRATES

PHYLUM	MOLLUSCA (100,000 species)	ANNELIDS (15,000 species)	ARTHROPODA (1 million species)
Sample organisms	Squid, Octopus, Clam, Snail	Earthworm, Leech	Insects, Spiders, Crustaceans
Circulation	Open Circulation	Closed Circulation	Open Circulation
Gas exchange	Gills and Mantle cavity	Gas exchange by diffusion through skin, oxygen in vessels	In insects and spiders tracheae, in crustaceans gills are used
Excretion	Nephridium	In every segment is a pair of nephridia.	Insects have malphigian tubes
Nervous sytem	Simple sense organs,and 3 pairs of ganglia	Simple brain,a pair of nerve cords, simple sense organs	Brain, a pair of nerve cords, developed sense organs
Reproduction	Sexual, separate sexes, mating in water	Sexual, hermaphroditic but mating happens.	Sexual, separate sexes
Support and Motion	Most have a hydrostatic skeleton, move with feet	Hydrostatic skeleton, move with developed muscles	Intact outer skeleton, developed muscles and jointed appendages
Environment and life style	Mostly aquatic, some terrestrial; carnivores, herbivores, scavengers	Mostly aquatic, some terrestrial; carnivores, herbivores, scavengers	Mostly aquatic, some terrestrial, carnivores; herbivores, scavengers
Other features	Mostly shelled but soft-bodied	Earthworms dig into land and aerate it.	They have a strong skeleton.

Examples of parasitic nematodes (in humans) are pinworm, hookworm, ascaid worms, trichinae worms, and whipworm.

7. Rotifera (Rotifers: Wheel animals)

Rotifers are microscopic multicellular organisms that live in water. They have cilia around the mouth. These cilia have a role in the feeding of the organism. Rotifers have a developed digestive cavity but they are pseudocoelemates. Nervous system consists of a brain and some sensory organs (eye spot). There are protonephridia for the excretion of excess water and wastes. Most live in freshwater. There are around 1500-2000 species.

8. Mollusca

About 50,000 living and 35,000 fossil species of molluscs are known. Examples of the group include mussels, octopuses, snails, slugs, oysters and squid (Figure 11.10). Many species are marine but there are some freshwater mussels and snails, and many terrestrial slugs and snails.

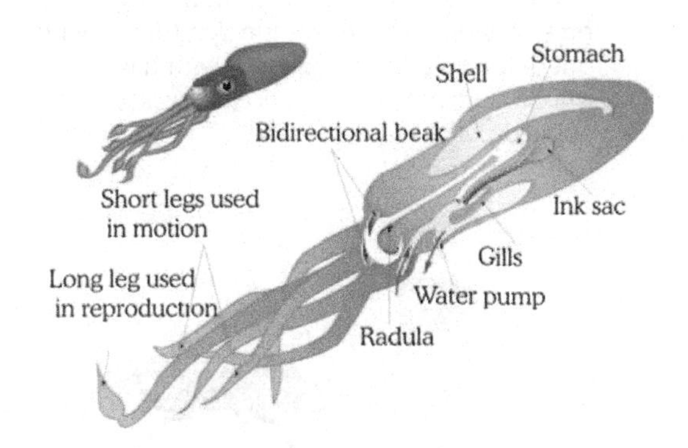

Figure 11.10.: Squid structure

Their soft bodies are usually covered dorsally with a hard shell made of $CaCO_3$. A flat, broad, muscular foot is used for locomotion. Mantle and mantle cavity carry shell-producing glands. Digestive system consists of a mouth buccal cavity, esophagus, stomach, intestine and anus. Many species have an open circulatory system and gills (Figure 11.11).

SEGMENTED WORMS

Lumbricus terrestris (Earthworm)

Digestive, excretory, circulatory and reproductive systems are developed. Gas exchange takes place by diffusion across skin cells. Mucus secreted by epidermal glandular cells provides protection and moisturization of the external surface. These animals have closed circulatory systems and hemoglobin in the blood (thus the blood is red).

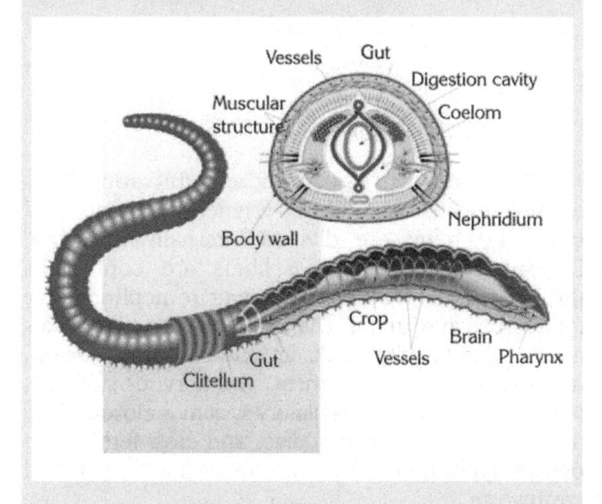

Hirudo medicinalis (Medicinal leech)

The body is equally segmented. They attach themselves to a vertebrate host and are able to suck up to 15 gr. of blood. While feeding, hirudin, an anticoagulant secreted by leeches, ensures the continuation of the blood flow. Leeches are found mostly in muddy waters and streams.

Leeches are used to remove blood accumulations, but this may be dangerous because they are vectors of tuberculosis and typhus. Diseases may be transferred from one person to another by treatment.

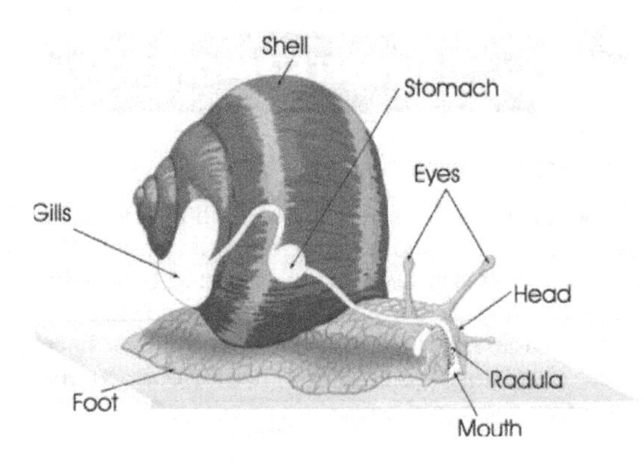

Figure 11.11.: Snail structure

9. Annelida (Annelids)

There are about 15,000 species in this group. They are segmented, with over 100 segments in exceptional cases. Segments are divided internally by septa. Digestive tract and nerve fibers are continuous throughout the segments. One pair of nephridia are found in each segment. The segmented body provides agility, and elasticity results from the separate coelom and muscles of each segment. The nervous system is ladder-like, while the circulatory system is closed. They are completely hermaphroditic, and cross-fertilization occurs. Typical examples are medicinal leech and earthworm.

Figure 11.12.: Earthworms (Annelids) increase soil productivity in cultivated areas.

Preservation of animal samples

Collected material may be kept alive or preserved. In general, collected samples are preserved. In order to keep the animals for a long time, the following methods are used:

- One method is fixation, in which the aim is to prevent decay or corruption. Animals are kept in 70% alcohol or 4.5% formaldehyde solutions for this purpose.
- Some invertebrates like insects may be kept in a dry condition (by pinning). Antennae, legs, and wings may be streched before drying. Butterflies are pinned with an insect needle (or a stainless pin) through the thorax. They are then stretched and held in this position for a while (1-2 days) with the help of pins. In this way insects can be preserved for long periods.
- Birds and small mammals are preserved for different purposes. One common method of preservation is to remove the skin (or fur) without damaging it and stuffing it with hay or cotton. To prevent decay, naphtaline should be applied periodically.

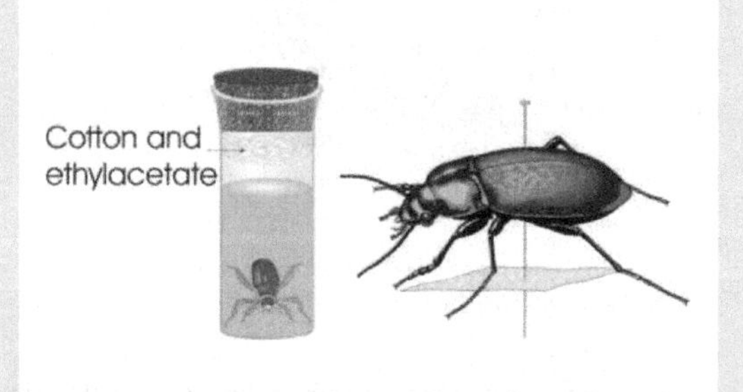

- All collected materials must be labelled. A label usually includes the common and scientific names, family name, habitat, date of collection, and name of the collector.

10. Arthropoda

There are about 1,000,000 species of arthropods found in all habitats. Arthro means joint, pod means foot. The body is segmented with the segments usually arranged in groups to form a head, thorax and abdomen. The skin is covered by a cuticle made of chitin polysaccharides of epidermal origin. This is called an exoskeleton. The exoskeleton is armor-like in crustaceans in which $CaCO_3$ is a component (Figure 11.13).

Figure 11.13.:Some arthropods have hard exoskeletons containing $CaCO_3$.

The exoskeleton must be shed (molting) 4-7 times during growth.

a. General characteristics of arthropods:

- Bilateral symmetry.
- Jointed legs develop on each segment (or just on thoracic segments).
- Exoskeleton is chitinous and molting occurs.
- Striated muscles are present, which provide agility.
- Digestive system is complete; mouth structures differ according to diet.
- Open circulatory system.
- Respiration takes place through trachea, book lungs, gills and skin.
- Malpighian tubules are the excretory organs.
- Nerve system is ladder-like.
- There are separate sexes. Direct development or metamorphosis are seen.

There are four classes of arthropods:

- Crustacea (crustaceans)
- Arachnida (spider and allies)
- Insecta (Insects)
- Chilopoda - Diplopoda (Centipedes and Millipedes)

b. Crustacea

They live in seas, freshwater and on land. Examples are crab, shrimp, lobster and Daphnia (Figure 11.14).

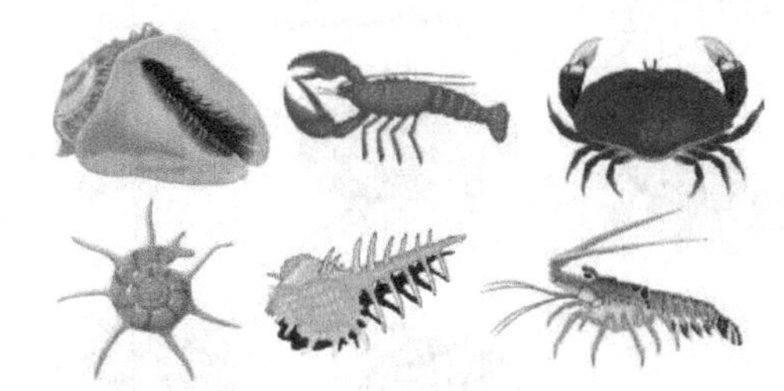

Figure 11.14.:Some examples of crustaceans

Their soft bodies are usually covered dorsally with a hard shell made of $CaCO_3$.

c. Arachnida

Body consists of a cephalothorax and abdomen as in the crustaceans. In some groups the cephalothorax and abdomen are fused together. All are terrestrial. Examples are scorpion, spider, tick and scabies mite. (Figure 11.15)

Figure 11.15.: Some species of scorpions and spiders

d. Insecta

Insects, the largest terrestrial animal group, are common throughout the world. The insect body is divided into head, thorax and abdomen. A pair of antennae, compound eyes, simple eyes of varying numbers and 3 pairs of mouth parts are found on the head (Figure 11.16-17).

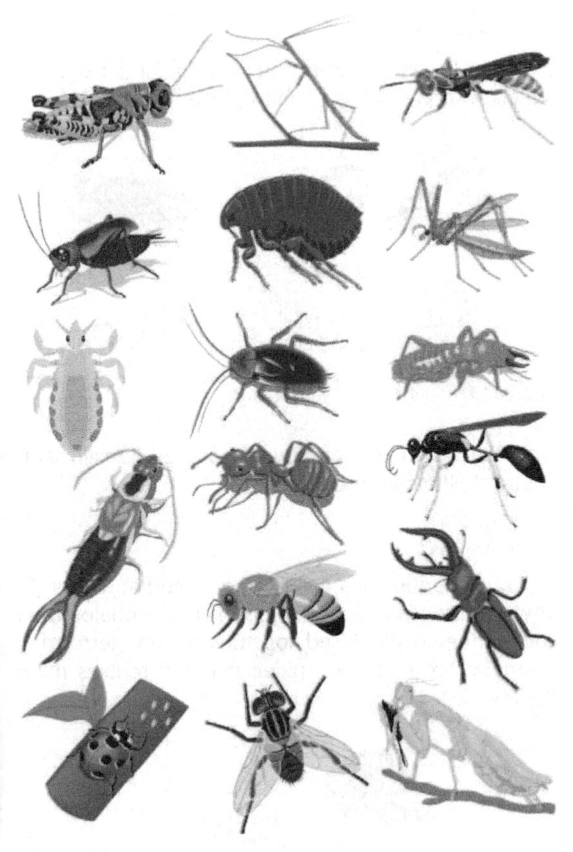

Figure 11.16.: Insect types

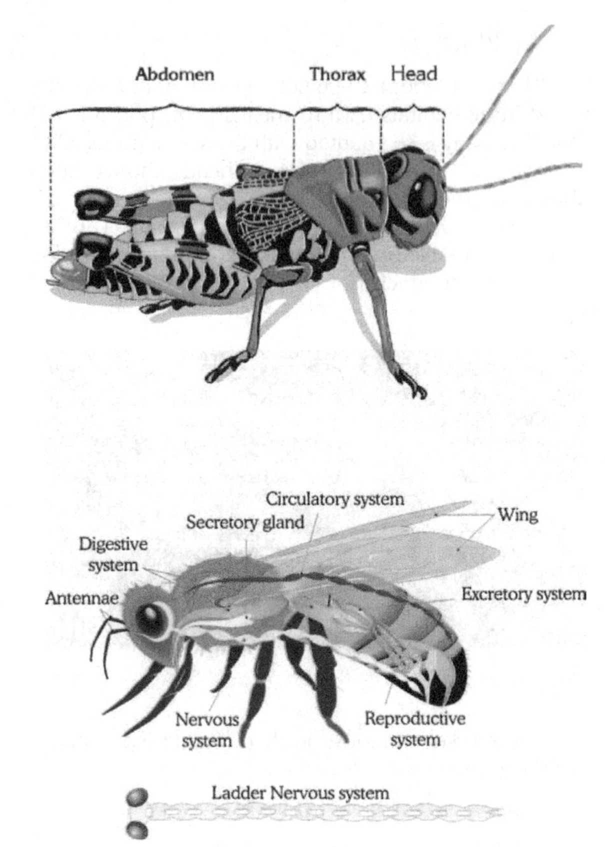

Figure 11.17.: General structure of insects

e. Chilopoda – Diplopoda

All are terrestrial, but restricted to moist places. The segmented body consists of a head and trunk. Body segments have 1-2 pairs of legs. Millipedes are detritus feeders, while centipedes are predators(Figure 11.18).

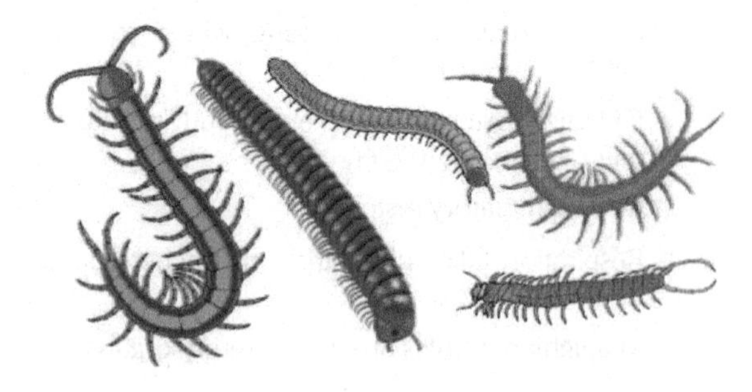

Figure 11.18.: Some types of centipedes and millipedes

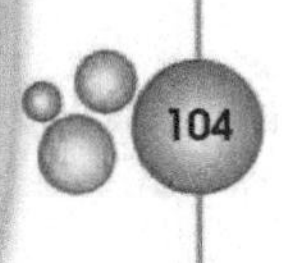

11. Echinodermata (Spiny-skinned)

Figure 11.19.: Some echinoderms form beautiful scenes in oceans.

These are the first deuterostomes, the most advanced group, consisting of only echinoderms and chordates. All are marine. There are 7000 living and 13,000 fossil species known. They are radially symmetric (secondarily) and have neither head nor segmentation. The coelom, coated with ciliated epithelia, is well-developed. The digestive system includes a mouth and anus. Respiratory organs are small gills. There is no heart and an open circulatory system is observed. The nervous system is composed of nerve rings with radiating nerves in the center. They have separate sexes. Asexual reproduction is by regeneration. Metamorphosis is observed during development. Examples are sea star, brittle star, sea cucumber, sea urchin and sea lily (Figure 11.20).

Figure 11.20.: Sea stars one of the best known echinoderms.

12. Chordata (Chordates)

Chordates have 3 embryonic layers (ectoderm, endoderm and mesoderm) and are bilaterally symmetric.

a. Chordate characteristics.

- A more or less developed dorsal endoskeleton and notochord is present. The notochord is permanent in some groups, temporary in others, while in some it develops into vertebrae.
- A brain develops as an extension of the anterior tip of the notochord. The rest remains cylindical and forms the spinal cord. Nerves protruding from both spread throughout the body.
- Ventral to the notochord lies the digestive tract. The anterior end of the tract is specialized into respiratory organs. Aquatic species have gills but terrestrial ones have lungs. Heart is positioned ventrally.

Chordates are classified into three groups:

b. Urochordata (Tunicates)

All are marine. Some are solitary and free-living but become sessile after a free-swimming larval stage. There are colonial species, too. Notochord and nerve cord are found only in the caudal part of larvae. Adults have only a small node as a remainder of the nerve cord.

c. Cephalochordata (Lancelets)

In lancelets, the head is absent. Notochord is present in adults and functions as a skeleton. An example is Amphioxus lanceolatum (Figure 11.21).

Amphioxus species are littoral, small (5-6 cm long), translucent yellow animals. Both ends are pointed. To the anterior part of the digestive tract there are gill slits in pairs. In the area between the nerve cord and gut, is the notochord, which persists in the adult.

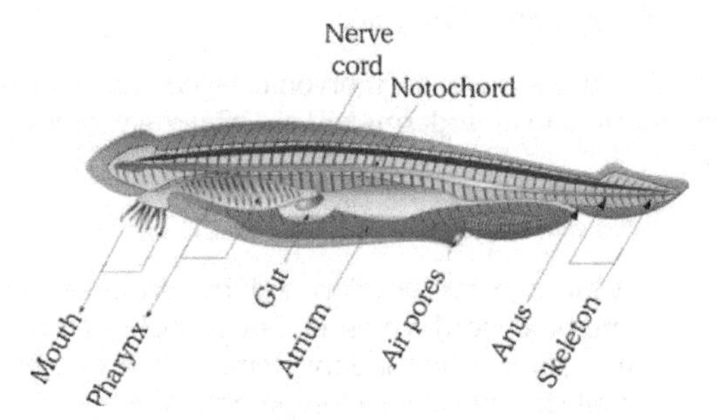

Figure 11.21.: Structure of Amphioxus. They are intermediates between invertebrates and vertebrates.

Amphioxus (or lancelet) species resemble both invertebrates and vertebrates. Single-layered epitheilum, absence of a heart, colorless blood, and protonephridia as urinary organs are invertebrate characteristics, whereas presence of notochord and nerve cord, presence of a respiratory organ anterior to the digestive tract, and pattern of blood flow are typical of vertebrates.

d. Vertebrata (Vertebrates)

All vertebrates have developed brains and skulls (crania). The notochord is observed only in the embryo, in the fetus and adult being replaced with vertebrae. Vertabrates have an endoskeletal structure.

d.1. Vertebrate Characteristics

- Epidermis is multi-layered.
- Skeleton is jointed, either cartilaginous (in sharks and jawless fishes) or bony (all others).
- There are two pairs of extremites. These are fins in aquatic species and limbs in terrestrial ones. Joints attaching limbs to the vertebral column are at the scapular arch and the sacral arch. Skeletal muscles function in movement.
- Digestive tract begins at the mouth, which opens to a stomach, followed by intestines, and ends in the anus. There are digestive glands, liver and pancreas that aid digestion.
- Circulatory system is closed. Hemoglobin is the pigment that carries CO_2 and O_2 in the blood. Blood is red. Heart contains 2 to 4 compartments (atria and ventricles).
- Kidneys are the urinary organs.
- Most have separate sexes. Paired gonads produce germ cells released from an opening near the anus. Metabolites, filtered by kidneys, and germ cells are carried through a common channel. Because of this, the system is called the urogenital system.

There are 2 groups and 6 classes:

Anamniota (These lack an amniotic sac)

Class: Agnatha (Jawless fishes)

Class: Pisces (Fishes)

Class: Amphibia (Amphibians)

Amniota

Class Reptilia (Reptiles)

Class Aves (Birds)

Class Mammalia (Mammals)

Except for birds and mammals, all are cold-blooded and depend on the environment for heat. Homeotherms (birds and mammals) can stabilize their own body temperature internally.

d.2. Agnatha (cyclostomata, jawless fishes)

These have eel-like, cylindical bodies with no jaw, paired fins or scales. They are the simplest vertebrates and retain a notochord throughout life. They are parasites and scavengers. They are found in freshwater and seas.

d.3. Pisces (Fishes)

Cartilaginous (Chondrichthyes) and bony (Osteichthyes) fishes are the two classes in this group. The skeleton may be cartilaginous (shark) or bony (many fishes, e.g. carp) . Epidermis may have scales or not (Figure 11.22).

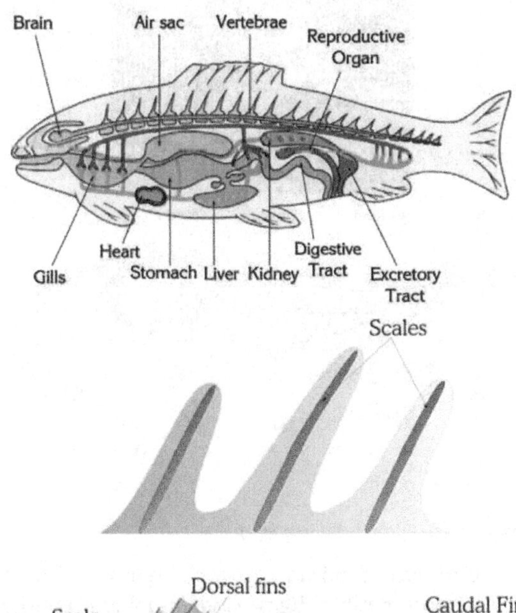

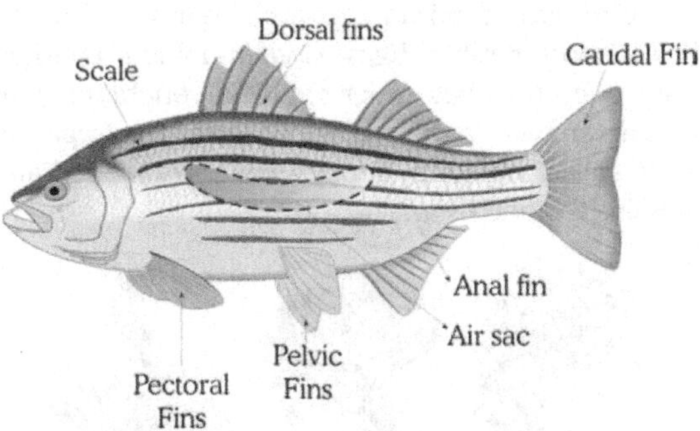

Figure 11.22.: Fish anatomy and morphology. Identification and classification is made using morphological data like scale structure and arrangement.

Gills are the respiratory organs. In cartilaginous fishes, there are 5-7 gill pairs, but 4 in bony fishes. Additionally gill pairs in bony fishes are covered with a bony flap, the operculum.

In fish, the eyes are focused at a shorter distance while resting and are able to see 1 m at most. Focusing is done by the backward and forward movement of the lenses. Eyelids are rare.

Vertebrae are variable in number (between 70-400). Ribs are unattached at the ends. Sexes are separate. Eggs are released and fertilization is external, except in sharks. Some are ovoviviparaous (e.g. Poecilidae) in which the eggs hatch one by one as the female lays them. The swim bladder functions in buoyancy, and fins are used for swimming. Adults have mesonephros and lack urinary bladders.

Eels, trouts, mackerels are a few examples.

d.4. Amphibia (Frogs and Salamanders)

Anurans (frogs) live both in water and on land. Amphibia means double life. Amphibians either live entirely in water or, usually, they return to the water to reproduce. Mucus glands and poision glands are found under the skin. Skin secretions protect the body from bacteria and retain moisture. Respiration occurs partially through the skin. Eggs hatch in water. Tadpoles (larvae) are tailed and respire with gills (some salamanders remain in this form). Metamorphosis follows and larvae become adults. Adult frogs lose their gills, tails and caudal fins. They respire with lungs and their heart has 1 atrium and 2 ventricles (Figure 11.23).

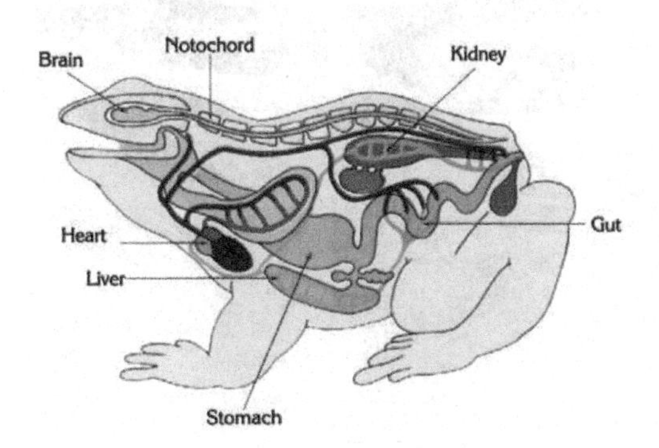

Figure 11.23.: Internal anatomy of frogs.They are interesting because they can live both in water and on land.

d.5. Reptilia (Reptiles)

Many reptiles are terrestrial. Their thick, dry skin is covered with scales and plates. They have some glands in their skin. Ears have columella as in birds and frogs. Limbs have 5 digits, but in snakes and in some lizards some or all limbs are missing. A reptile's internal organs are contained within a rib cage. The heart and lungs are well-developed. Internal fertilization is a remarkable feature of reptiles (Figure 11.25-26).

WHICH SNAKE IS POISONOUS?

The majority of snakes are harmless. These are identified by thinner heads, longer and thick necks, whip-like tails, 1 pair of anal plates and large scaled head. In poisonous snakes head shape is triangular neck is short and narrow, pupils are vertical or oval-horizontal, tail is truncate and short anal plate is single and there are 2 fangs inside mouth. Snakes jaws are jointed elastically so that they can open their mouth very wide. Each jaw has hook-like teeth (Fig-11.24).

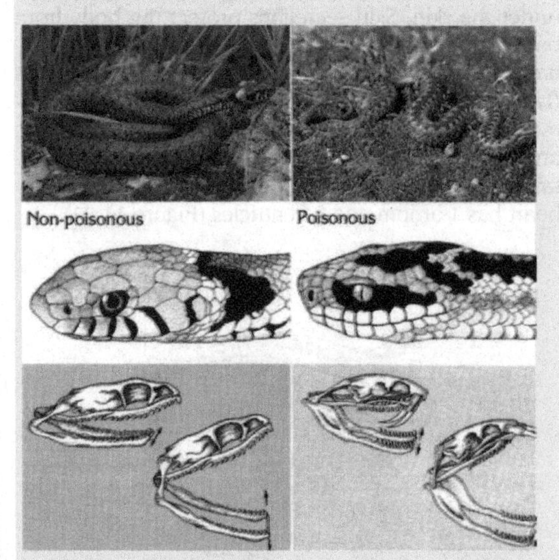

Figure 11.24.:Structure in snakes. Their head structures are different in poisonous and harmless species.

Usually snakes eat mice, rats and voles. Accordingly, they are useful in agriculture. But some poisonous species are able to kill humans and pets. Poison is dark, clear, yellowish or colorless. Entering the body, poison blocks the nervous system and respiratory functions. Furthermore some enzymes inside poison agglutinate blood cells and harm blood tissue. Serum obtained from horses is used in therapy. The most effective action after bite would be to remove poison from the tissue as soon as possible. For this, the area a bit higher than the wound is constricted with a rope (blocking the flow), then the wound should be cut and blood with poison must be removed by spitting.

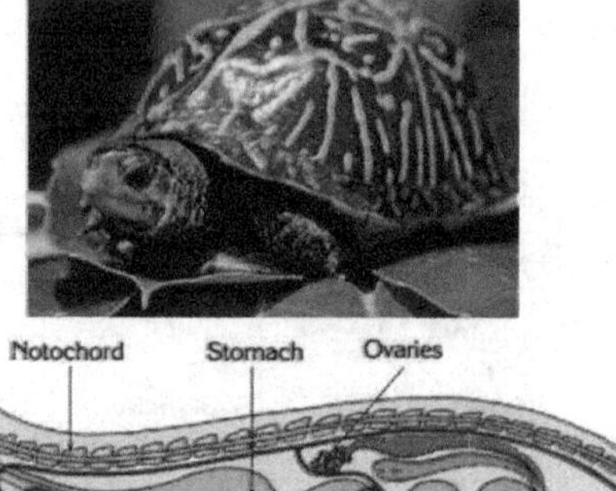

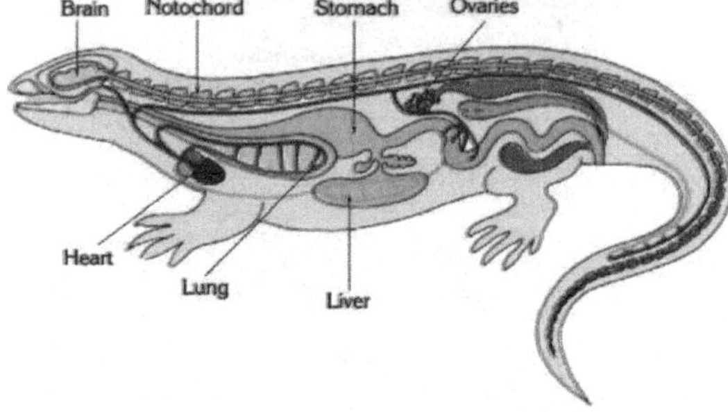

Figure 11.25.: Internal anatomy of reptiles

As in birds, reptilian eggs are rich in yolk, but the shell is more flexible. Eggs contain amnion, chorion, allantois and vitellus structures. The vertebrates that have an amnion in the egg are collected in a separate group called amnionts. Reptiles are cold-blooded animals like fish and frogs.

Figure 11.26.:Reptile types

d.6. Aves (Birds)

The bird body is remarkably covered with feathers. Feathers provide insulation and prevent water loss, and function in flight as well. Hollow bones and air sacs also facilitate flight. Their tongues are hard and they have bills without teeth. Their anterior extremites are wings which function in flight. A rib cage protects internal organs (Figure 11.27-28)

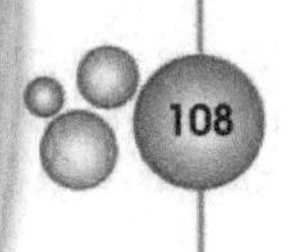

Figure 11.27.: A bird's organs are specialized for flight.

Birds are warm-blooded. Their heart consists of two atria and two ventricles. Respiration is well-developed. Development and reproduction is similar to that of reptiles. Eyelids are movable. There are upper, lower and inner eyelids. The iris shrinks or enlarges to focus. Kidneys are of metanephric. There is no urinary bladder. Urine and feces are expelled through the cloaca. Because the kidney glomeruli are very small, only a small amount of fluid can pass from the blood to the kidneys. Most is reabsorbed. In this way, water loss is kept to a minimum.

Because they can fly, distribution of birds is wider than other terrestrial vertebrates. Birds show great diversity. The bill, foot, wing and tail are highly variable and adaptable organs.

Figure 11.28.: Types of birds

d.7. Mammalia (Mammals)

Mammary glands (in females) and hair covered bodies are characteristics of mammals. Other features include (Figure 11.29):

- Hair covers the skin. Hair originates from the epidermis, while hair follicles are in the dermis.
- They are warm-blooded and their hearts have 2 atria and 2 ventricles. Except camels, the erythrocytes lack nuclei.
- Coelom is divided by the diaphragm into thoracic and abdominal parts.
- Teeth are developed.
- Respiratory system is well-developed.
- Most have a placenta and give birth to live young.
- Brain function and mobility are developed.

According to organization and embryonic development, there are 3 subclasses:

- **Monotremata (Egg-laying mammals):** These organisms do not have a placenta. The urogenital opening is a cloaca into which the large intestine, and urinary and genital ducts open. This feature is shared with reptiles and birds. These are egg-laying animals, but they feed their young with milk produced by mammary glands. There are many extinct species and a few living species. Examples are the duck-billed platypus (Ornithorhynhus anatonus) and the spiny anteater (Echidna aculeata).
- **Marsupialia (Pouched Mammals):** The pouched mammals do not have a placenta. They give birth to underdeveloped embryos which complete their development inside a pouch (marsupium) on the mother's abdomen, where the mammary glands are located. Many species are found exclusively in America and Australia. They are hunters or herbivores. Examples of the group are the kangaroo, sugar glider, wombat, and opossum.
- **Placentalia (Placental mammals):** Nearly 95% of mammal species found today are in this group. The placenta plays an important role in embryonic development, functioning in excretion, respiration, and the transfer of nutrients from the mother to the embryo.

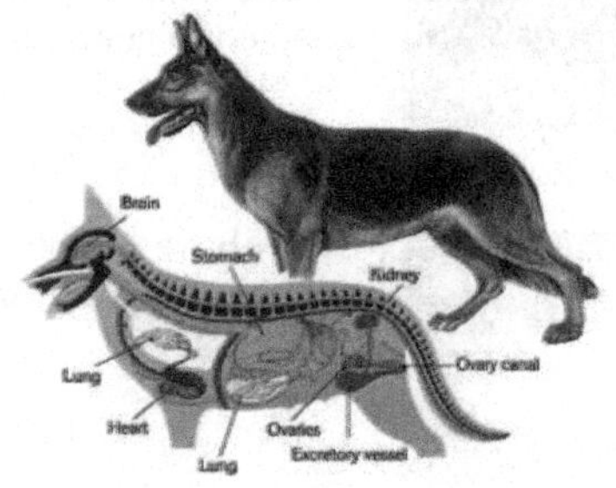

Figure 11.29.: Internal anatomy of a dog , a type of mammal.

PLACENTAL MAMMALS

Insectivora (Moles, shrews)

These are generally nocturnal animals with thick fur, protruding proboscis, and sharp teeth. Some are arboreal while many others are subterranean.

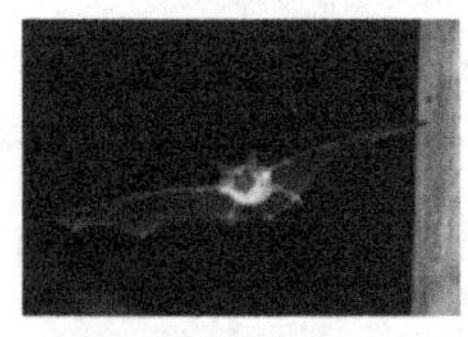

Chiroptera (Bats)

These are mostly insectivorous animals that are able to fly. Some species suck blood from vertebrates and some others prey on small vertebrates like frogs. They are usually nocturnal.

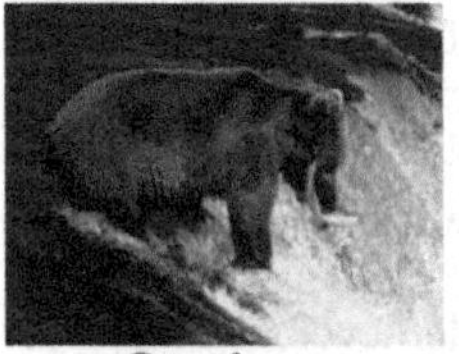

Carnivora (carnivores)

They eat basically meat and are characterized by sharp and long canine teeth. The front molars are sharp, too. They are strong and fast predators.

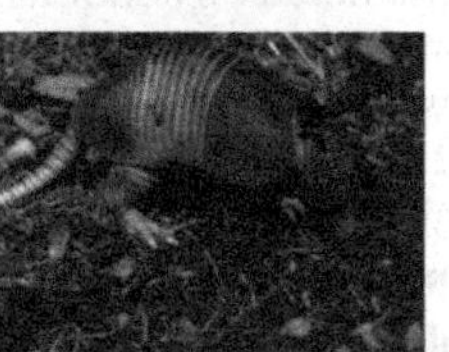

Edentata (Toothless mammals)

They are sluggish animals that feed on insects and small invertebrates. Teeth are reduced or absent. Body is protected by thick plates.

Rodentia (Rodents)

Incisors are characteristically well-developed. Canine teeth are missing. They are adapted to gnawing.

Lagomorpha (Hares and rabbits)

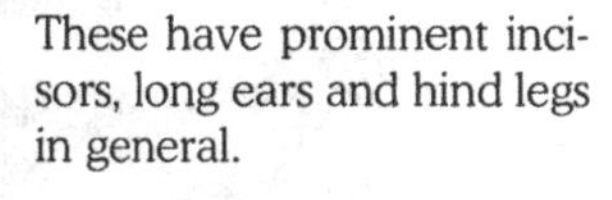

These have prominent incisors, long ears and hind legs in general.

Primates (monkeys, chimpanzee)

The brain and eyes are well developed. Orbits are directed anteriorly. Some have nails instead of claws. They are basically omnivorous.

Perissodactyla (horse, zebra)

These are herbivores with an odd number of hoofed digits (1-3-5, generally 1). Teeth are specialized for grinding of plant matter. They are normally robust, long-legged plains dwellers.

Artiodactyla (cattle, sheep, giraffe, pig, deer)

These are herbivores with 2 or 4 hoofed digits per leg. Many are ruminants with chambered stomachs. Horns are usually seen in one or both sexes.

Proboscidea (Elephants)

These are the largest terrestrial organisms (up to 7 tons). A heavy head, large ears, loose, thick skin, and muscular body are characteristic of the group.

Cetacea (Whales and Dolphins)

Sea mammals are included in this group, despite rare occasions of freshwater species. All extremites are adapted to sea life. Hind limbs are missing. Many species have a thick layer of fat. They are among the most intelligent animals. The blue whale is the largest animal ever to have existed in the world.

Pinnipedia (Sea lions and seals)

Limbs are specialized for swimming. They are carnivores and feed primarily on fish.

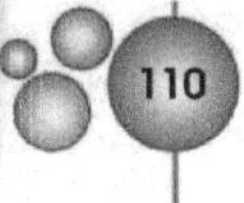

SELF CHECK

ANIMALS

A. Key Terms

Porifera	Turbellaria
Enidaria	Trematoda
Polyp	Cestoda
Medusa	Nemertea
Hydrozoa	Nematoda
Scyphozoa	Rotifera
Anthozoa	Annelida
Platyhelmintes	Insecta
Echinodermata	Chordata
Pisces	Reptilia
Amphibia	Aves
Mammoloid	Monotremata

B. Review Questions

1. For centuries sponges were classified as plants. Justify their current classification as animals.
2. Why is the sea a more hospitable environment for many animals than the land or fresh water?
3. What are the advantages of bilateral symmetry and cephalization?
4. Describe how hookworms and tapeworms affect people.
5. Describe the circulatory and digestive systems of an earthworm.
6. How do the sense organs of a planarian and an earthworm differ?
7. Explain how planarians and earthworms are adapted to their environments.
8. What are the main body features of all arthropods?
9. Describe the body of a spider.
10. Compare a crayfish's eye with a spider's eye.
11. What do you think is the main disadvantage of an exoskeleton? Why do you think so?
12. What special characteristics do all insects have in common?
13. Give several reasons why insects are able to live in so many different environments.
14. What features do all echinoderms have in common?
15. Compare the two ways that starfish reproduce.
16. What are the three characteristics of chordates?
17. How are fish related to invertebrates?
18. Give two reasons why an amphibian cannot spend its entire life away from water. Describe how reptiles can do this.
19. Explain why fish and amphibians lay so many eggs.
20. The chemical DDT was once sprayed into swamps to kill mosquitoes. Later it was found to cause birds to lay eggs with very thin shells. Why would this be harmful to the birds? Support your answer.
21. Life scientists noticed that deer in northern areas are larger than deer that live closer to the equator. Suggest a reason why the deer are different sizes.
22. Give three reasons why a frog and a mouse are classified in the same phylum.
23. Give three structural differences between a frog and a lizard.
24. State three features of birds that have contributed to the success of this group.

C. Matching

a. Asymmetrical	( ) intermediate layer
b. Coelom	( ) carries both male and female gametes
c. Hermophrodite	( ) No equal places
d. Mesoderm	( ) Sessile cnidaria
e. Polyp	( ) Body cavity
f. Medusa	( ) Swimming cnidaria

D. True or False

1. ☐ Fishes respire by their gills.
2. ☐ Nematods have open circulatory system.
3. ☐ Spiders are member of erustacea.
4. ☐ Kangaroo is an egg-laying mammal.

E. Fill in the blank

1. Animals without backbones are properly referred to as ....
2. .... animals lack a body cavity.
3. Two body forms found among cnidarians are the......................... and the ....
4. .... symmetry is characteristic of cnidarians.
5. In sponges, a water current is created by the .... cells.
6. .... are mammals that lay eggs.
7. Animals that can maintain a constant body temperature are known as ....

F. Multiple choice

1. Which of the following aspect <u>does not</u> belong to crusteans

A) Respiration by gills B) Aquatic organisms C) Open circulatory system D) Hermophrodyte E) Metamorphosis

2. Which of the following has most primitive structure among the invertebrates?

A) Spongy B) Coelenterates C) Flatworm D) Arthropods E) Cnidaria

3. Which one has the simplest heart structure?

A) Mammal B) Frog C) Reptile D) Bird E) Grasshoper

4. Which of the following lays eggs out of its body?

A) Cat B) Eagle C) Dolphin D) Human E) Monkey

5. Which of the following is a warm-blooded animal?

A) Frog B) Lizard C) Crocodile D) Cat E) Snake

6. Which of the following <u>is not</u> property of arthropods?

A) Have an open circulatory system
B) Hard skeleton made of chitin and protein
C) Gas exchange occurs through body surface
D) Reproductive phase is composed egg , larvae , pupa and adult
E) All of them

7. Which of the following is correct for amphibia?

A) Heart is four chambered
B) More primitive than fish
C) Their body temparature is variable
D) They respire using gills in adult stage
E) All of them

8. Which of the following <u>is not</u> property of reptiles?

A) Respire using skin
B) Turtle , cobra and crocodile are reptiles
C) Reproduction by egg
D) Body temparature is variable
E) Three-chambered heart

9. In which of the following organism is seen metamorphose?

A) Insect-Frog B) Worms-Insects C)Frog-Fish D)Amoeba- Frog E) Insect-Fish

10. Which respiration <u>is not</u> made in the following organisms?

A) Frog → Lung and skin
B) Eartworm → Skin
C) Spongy → Body surface
D) Crocodile → Gills
E) Fish → Gills

11. Jelly fish and sea anemones are examples of ....

A) flatworm B) coelenterates C) plants D) roundworms E) mollusca

12. Which of the following has four chambered heart?

A) Snake B) Lizard C) Fish D) Crocodile E) Frog

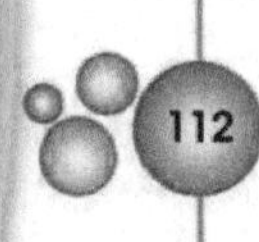

Made in United States
Orlando, FL
26 December 2024